T0155854

The Chernobyl, Fukushima Daiichi and Deepwater Horizon Disasters from a Natural Science and Humanities Perspective

Volker Hoensch

The Chernobyl, Fukushima Daiichi and Deepwater Horizon Disasters from a Natural Science and Humanities Perspective

 Springer

Volker Hoensch
Penzberg, Bayern, Germany

ISBN 978-3-662-65318-0   ISBN 978-3-662-65319-7   (eBook)
https://doi.org/10.1007/978-3-662-65319-7

This Springer imprint is published by the registered company Springer-Verlag GmbH, DE part of Springer Nature.
The registered company address is: Heidelberger Platz 3, 14197 Berlin, Germany

Technical disasters are
socially responsible.
Natural events against which there is no
or insufficient precautions have been taken
are not to be accepted as fate

For my wife Elisabeth

Preface

One could well get the impression that we are living in a time in which catastrophic events are increasingly burdening our social coexistence.

We decided that we would focus on four individual events with regard to this impression.

At the beginning we put the ballad "The Sorcerer's Apprentice" by Johann Wolfgang von Goethe. The sorcerer's apprentice is alone and tries a spell of his master to prepare a bath for himself. By his action, the sorcerer's apprentice exposes himself and the community to what the sorcerer's apprentice sees as a manageable risk. An examination of the concept of risk is therefore inevitable. With the appearance of the sorcerer's apprentice, the "drowning" of the building is averted. The risk, initiated by the sorcerer's apprentice, was averted by the warlock, it was mastered. The ballad "The Sorcerer's Apprentice" is characterized by the structure of the sequence of actions.

Is the structure formed by the action also in the three catastrophic individual events of

- Chernobyl (26 April 1986, explosion of reactor 4),
- Fukushima Daiichi (11 March 2011, destruction of several power plant units by a tsunami wave)
- Explosion of the Deepwater Horizon oil rig in the Gulf of Mexico (20 April 2010)

available?

These three individual events and the ballad "The Sorcerer's Apprentice" are analyzed in terms of their action features.

In the case of the ballad "The Sorcerer's Apprentice" and Chernobyl, as well as Deepwater Horizon, human actions are evident as having triggered the accident. But there are also opinions, especially in the case of Chernobyl, that assume a technical failure. This is the contradiction we are going to resolve. The Fukushima Daiichi disaster is also seen, on a superficial view, as a natural event. In contrast, it is due to human decisions, brought about by the selection of the site for the power plant and an inadequate state of protection against external and internal accidents.

This assessment becomes clear when the four accident sequences are compared with the so-called Swiss cheese model.

In the second chapter, we seek an answer to the question of direction for the four individual events.

The fact that a cause is followed by an effect with a temporal and spatial distance is an everyday experience that describes the problem of causality. Nature seems to have an inherent arrow of time that is not known to the basic laws of physics. A variety of different arrows of time are discussed in science. We will focus, as Stephen W. Hawking did in Chap. 2, Hawking (1994), on the thermodynamic, the psychological, and the cosmological arrows of time, emphasizing the special role of the thermodynamic arrow of time, which represents the growth of disorder or entropy. This allows us to distinguish past and future; time is given direction. This phenomenon can be read sociologically as a law of increasingly rational action orientation. Human cognition, that is, the perception, recognition, and processing of information, creates the foundations of the intentional structure necessary for a task to be accomplished through purposeful action. Intentions to act, that is, reasons and purposes, are not causes as highlighted by the application of the cause-effect structure.

Jens Rasmussen has developed a model for the cognitive demands placed on humans by the processes of information processing. He distinguishes three levels: skill-based, rule-based, and knowledge-based. In making decisions in everyday life situations, people may choose shortcuts between these levels. Rasmussen has developed what he calls a stepladder model for such shortcuts. This stepladder model allows associative jumps between all decision levels, and thus freedom of action. For an intentional structure shaped in this way, a heuristic is presented that is commonly referred to as the Rubicon model. Once the Rubicon is crossed, there is no going back in the intentional structure of action. The heuristic of the Rubicon model is projected into the cause-effect structure. Under the aspect of the causal principle, the metaphor of shooting with a bow and arrow is presented.

The causal chain in archery consists of the following links:

- Preliminary phase
- Cause, generation of an internal state
- Entry of an external system
- Point of no return
- Triggering event, causal principle
- Probabilistic influencing factors
- Effect

The archer performs the bow shot according to his intention, the hit "into the bull's eye."

The causal chain and the intentional structure are also explained for the ballad "The Sorcerer's Apprentice." The two nuclear catastrophes of Chernobyl and Fukushima Daiichi are described in the same way. The two structures are also elaborated for the explosion of the oil rig in the Gulf of Mexico, Deepwater Horizon. All four individual events are condensed by the "consequential reasons for action" (Nida-Rümelin).

Consequential reasons for action are directed at causally intervening in the world and generating a state of affairs that is different from alternative states of affairs. Consequentialism does not use the scientific concept of causality and apply it to action, but proceeds the other way round. Causality is assigned to human action. Or in other words: The intentional structure is determined by decisions.

In all four individual events, there is no convergence between the consequential reason for action and the intentional goal of action. The decision makers have spread out the dual character of the action.

In his decisions, man is subject to the processes of nature, whether he knows them or not. He must obey them. In the three individual catastrophic events, nature prevailed and society was burdened with the consequences.

Our decisions move within the spatiotemporal relational framework of physics and our consciousness. Experienced time connects consciousness and intentionality. Experienced time is subjective time, is consciousness of the present, the past, and the future. At the same time, we know that only the moment is real, the past is already past, and the future has not yet occurred. This raises the question whether there is also a basis in the circulation of nature for the difference between past and future. This difference goes beyond the differentiation between "earlier" and "later." It can only be grasped with a dynamic approach to the processes of being and becoming. These processes are viewed in the cognitive sciences as taking in information from the environment and the behavioral dynamics it triggers. We are talking about intentionality.

The three catastrophic individual events are placed in a spatiotemporal relational framework. The spatiotemporal relational framework is formed by the physical variables in order to be able to describe decisions and the resulting actions.

Now the arc of Chernobyl, Fukushima Daiichi, and the Deepwater Horizon drilling rig is being extended. Included is US Airways Flight 1549 from New York. The captain of the flight steered the plane onto the Hudson River and thus avoided a catastrophe. This presents the tension that shapes our worldview with regard to human decisions. Humans shape our view of the world through their capacity for consciousness; they acquire a creative power with which they must deal responsibly.

In the fourth and last chapter, we take up the conflict between the natural sciences and the humanities. The natural science side asks: How can there be reasonable causes in a world of causes? The humanities side asks: How can there be causes in a world of reasonable causes?

These two questions are answered by explanations of the cause-effect structure and the intentional structure. Processes by which the physical external world passes into the world of daily life familiar to consciousness lie outside the realm of physical laws. As a physical system, the brain is also subject to probabilistic laws, in addition to physical laws.

Decisions presuppose previous events and anticipate the unknown future. Decisions for an alternative course of action, this everyday challenge, belong to the Very Smallest. The structure of the Very Smallest is formed by the culture of a company, in short, corporate culture, the decision premises, and the decision processes. Corporate culture and decisions, formed by decision premises and processes, are the two sides of the same coin, "operational organization." When decisions are made in operational organizations, the dual nature of actions also comes into play. As we already know it. Every decision has a favoring and a burdening character.

The Very Smallest is embedded in throughout the whole book.

The Very Largest of all are the four elementary forces that arose from the original elementary force: Gravitation, the queen among the elementary forces, the Electromagnetic Force, the Weak Nuclear Force, and the Strong Nuclear Force. All four still control the processes in the universe today. Thus, we are immediately before an answer to the question of the direction.

Consequential reasons for action are the means of directing. The stage on which we humans act is determined scientifically by time, space, and causality, and sociologically by the community – we have limited ourselves to the operational enterprise. The script is written by the laws of nature. We have introduced entropy and the thermodynamic arrow of time derived from it as the "script writer." The arrow of time determines the development that humans try to influence with their decisions. These are the components that shape our worldview. Space, time, and causality are not objects. Objects of all kinds are restricted, finite, and conditional. The same is not true for space, time, and causality. Rather, space, time, and causality are the three "vectors" that span our reality, the basis of all knowledge, the precondition of all objecthood. And because reality is so large and inexhaustible, it can only be based on just such a foundation, which is to be stabilized by holistic security research.

The author's aim with this book is to show how holistic security research can be used and applied purposefully against the resistance that still exists.

Penzberg, Germany   Volker Hoensch
Christmas 2018

Contents

List of Figures

List of Tables

Four Selected Accident Events

1

1.1   The Concept of Risk

We would like to start with a quote from the former German Minister for the Environment, Prof. Dr. Klaus Töpfer:

> The expansion of human possibilities through the use of technical aids, be it airplanes, power plants, oil rigs or the like, made the creation of wealth possible. Technology must remain a tool for improving human living conditions. It must be calculable and controllable so that forces are not unleashed that could also bring about the end of human civilization. Greek mythology already teaches us that blessings and curses lie close together when man strives beyond his natural powers. Prometheus was punished by the gods because he brought fire to mankind and made their lives easier, while at the same time giving them godlike powers. This ancient message is more topical than ever. It is necessary to use the enormous possibilities of technical progress for the benefit of mankind, without at the same time becoming an outlaw of the divine order of creation.

> There is no alternative to technical progress. Only with the help of technology can we maintain prosperity in the industrialised countries, improve the living conditions of people in the Third World and also overcome environmental problems. However, we know today that with the expansion of technical possibilities, the risks also increase. (Hauptmanns et al. 1987)

Examples include the accidents at the Chernobyl and Fukushima Daiichi nuclear power plants and the explosion on the Deepwater Horizon oil rig, which will be discussed in more detail.

Further in the foreword by Prof. Töpfer:

> Modern technologies are having a more profound and long-term impact than ever on our human society and on the natural environment. Many fear a momentum of its own that can no longer be controlled. Unreflective growth thinking and blind faith in progress are

© The Author(s), under exclusive license to Springer-Verlag GmbH, DE, part of Springer Nature 2022
V. Hoensch, *The Chernobyl, Fukushima Daiichi and Deepwater Horizon Disasters from a Natural Science and Humanities Perspective*,
https://doi.org/10.1007/978-3-662-65319-7_1

therefore no longer responsible. Instead, technical progress must always be examined for inappropriate risks and dubious benefits. (Hauptmanns et al. 1987)

This means that we cannot avoid defining the prevailing understanding of the term "risk". In colloquial language, the term "risk" is associated with danger, i.e. the possibility of suffering damage. In English, a distinction is made between "danger" and "hazard". Danger is the possible effect of harm or the state of being threatened by a source of danger. Hazard is a source of danger, a risk. Although this differentiation helps us conceptually, we must note that there are different views on the concept of risk in the various scientific disciplines – engineering, social science and social philosophy, business administration and law.

1.2   The Sorcerer's Apprentice

Because of this shortcoming, we would like to turn to the ballad "The Sorcerer's Apprentice" by Johann Wolfgang von Goethe, which was written in 1797, the so-called Ballad Year (Sorcerer's Apprentice 2018).

The sorcerer's apprentice is alone and tries out a spell cast by his master. He uses a spell to transform a broom into a servant who must carry water to prepare a bath. The ballad begins with the following verses (http://www.reelyredd.com):

That old sorcerer has vanished
And for once has gone away!
Spirits called by him, now banished,
My command shall soon obey.

Every step and saying
That he used, I know,
And with sprites obeying
My arts I will show.

This quotation and the rest of the text of the ballad bring us closer to the scientific concept of risk, which is particularly common in the insurance industry. There, risk is essentially measured according to the objective extent of damage and its probability of occurrence, which is always determined in detail.

For the probability of occurrence in the sorcerer's apprentice it says: "That old Sorcerer has vanished".

The master is absent, so the sorcerer's apprentice can become active.

The extent of the damage is described by the words: "How the water spills: How the water basins, brimming full the he fills! Stop now, here me! Amle measure: Of your treasure. We have gotten!"

Later, "Brood of hell, you're not mortal! Shall the entire house go under?"

So much for the ballad "The Sorcerer's Apprentice."

But the ballad "The Sorcerer's Apprentice" suggests another consideration. Does the sorcerer's apprentice have the necessary competence to act? Does the sorcerer's apprentice act reasonably?

Obviously, the sorcerer's apprentice overestimates his competence to act and thus his knowledge. For this we repeat from the first quotation:

Every step and saying
That he used, I know.
And with sprites obey
My arts I will show.

Second quote:

He returns, more water dragging!
Now I'll throw myself upon you!
Soon, O goblin, you'll be sagging.
Crash! The sharp axe has undone you.
What a good blow, truly!
There, he's split; I see.
Hope now rises newly.
And my breathing's free.

Woebetide me!
Both halves scurry
In a hurry.
Rise like towers
Threr beside me,
Help me, help, eternal powers!

The sorcerer's apprentice does not have the knowledge to conclude his original intention to act with a positive result, he also lacks the necessary knowledge and thus the competence to act to limit the damage.

With the quantification of "probability of occurrence" and "extent of damage", the risk can be estimated.

In its most general form, the measure of risk is understood to be the product of the probability of damage, related to a unit of time, and the damage impact of the consequence:

$$\text{Risk value} = \text{probability of damage} \times \text{impact of damage}$$

Further, the ballad shows us to distinguish between controllable risk and uncontrollable risk. The sorcerer's apprentice realizes that he cannot control the risk he has summoned and in desperation calls for help:

Sir, my need is score.
Sprits that I've cited
My commands ignore

The master, on the other hand, masters the scene through his knowledge and shows competence in action:

To the lonely
Corner, broom!
As a sprit
When he wills, your master only
Calls your, then'tis time to hear it

The sorcerer's apprentice does not have this knowledge and is desperate:

Ah, I see it, dear me, dear me.
Master's word I have forgotten!

Ah , the world with which the master
Makes the broom a broom once more!

We summarize the action sequence of the ballad in key words:

- Overestimating oneself, proving one's supposed ability, deliberately exceeding one's competence,
- Ignorance of one's own doubts,
- Power rush, achieving personal success,
- Fear of consequences,
- Desperate for control,
- Rescue by the Sorcerer.

Transformed to the product approach to risk value:

- Probability of damage: event deliberately brought about to confirm one's own competence.
- Damage impact: manageable and containable.
- Risk: The master's intervention neutralizes the challenge deliberately made by the sorcerer's apprentice.

The assessments of the sequence of actions and the product approach for the risk value made here for the Sorcerer's Apprentice are also to be taken up in each of the three catastrophes presented below and used as a standard of assessment.

The assessments of the action sequences for the total of four events considered are summarised in Table 1.1.

So much for the recourse to Goethe's ballad "The Sorcerer's Apprentice".

1.3   Dealing with Knowledge

Now we can turn to the question of what knowledge is and how it comes about. This question belongs to the fundamental questions of philosophy.

The definition of knowledge and thus action competence is important, according to the motto "define your terms", in order to avoid that different facts are understood under the same term.

The question of what exactly the "essence" of knowledge is, how knowledge actually arises and is ultimately translated into decisions and action, has remained without a binding answer to this day: Is knowledge, after all, rather the cognitive process itself in the form of a continuous construction of people and social systems? How does knowledge ultimately become action? What role do emotions, motivations, will, attitudes and values play on the one hand, and social relations and culture on the other?

Against the background of such questions pressing for clarification, knowledge is not the domain of one discipline alone.

The intelligent, efficient and responsible handling of knowledge is a major social challenge and thus ultimately also an individual competence. Is individual competence able to distinguish between controllable and non-controllable risk? Where is the limit of the danger threshold?

The simultaneous perception of harm, costs and benefits of technology is not uniform in society. Mostly there is no conception for the evaluation of probabilities of occurrence (otherwise nobody would play the lottery, because the probability for 6 right numbers is just under 1:14 million). The individual preconditions, which are shaped by the natural and social environment, by education and acquired ethical and political foundations, determine emotional assessments from a very different individual level of knowledge and information.

It can be said with certainty that the danger threshold was exceeded in the following three events:

* the accident at the Chernobyl nuclear power plant,
* the incident at the Fukushima Daiichi nuclear power plant, and
* the explosion of the oil rig "Deepwater Horizon",

which we would now like to discuss in more detail.

1.4   Chernobyl (26 April 1986; Explosion of Reactor 4)

A large amount of literature exists on this incident. We mainly rely on (Reason 1994), because there the technical accident sequence was extended by the human component.

The commissioning programme of a reactor also includes the experimental validation of the accident concept. The accident concept includes demonstrating that the no-load capacity of a turbine generator is sufficient to supply power to the emergency cooling system for the reactor core for a few minutes if a usable voltage generator is available. This would bridge the time until the diesel-powered backup generators are ready for use.

A voltage generator had been tested on two previous occasions but had failed due to a rapid voltage drop. On the 26 April 1986 test, the aim was to repeat the test before the reactor was due to be shut down for its annual inspection, which was imminent.

The experiment is characterized by the following chain of events:

On April 25, 1986, at 1:00 p.m., the reduction of reactor power begins with the aim of establishing the experimental conditions. The test was to be carried out at about 25% of the nominal reactor power (in the order of about 700 MW) in Unit 4. At 14:00 the emergency cooling system is disconnected from the primary circuit. At 14:05 the dispatcher from Kiev (supervisor for the power grid) orders to continue power generation of reactor 4. The emergency cooling system, which had previously been shut down, is not reconnected. At 23:10, reactor 4 is disconnected from the power grid. At 00:28, the reactor operator resumes the test. This fails to maintain reactor power, resulting in very low power. At this point the test should have been stopped given the very low power. The operator continues to attempt to control the reactor in an unknown and unstable area in order to continue the planned test, in the process the reactor exceeds the critical point. The overshoot is irreversible. The chain reaction gets out of control, at 01:24 the reactor explodes.

The chaos inside the damaged reactor under the sarcophagus and the pollution of the entire environment are unimaginable.

The main cause of the disaster is considered to be the design characteristics of the graphite-moderated nuclear reactor (type RBMK-1000; transcribed reactor Bolshoi Moshchnosti Kanalny, roughly high-power reactor), operation in an inadmissibly low power range and serious violations of applicable safety regulations by the operators during the test. The minimum value of the shutdown reactivity (reactivity is the measure of the deviation of a nuclear reactor from the critical state. The neutron multiplication factor k is the quotient of the number of neutrons produced divided by the number of neutrons absorbed and discharged. Instead of k, one often uses the "reactivity," ϱ; $\varrho = k - 1$ divided by k. The reactivity measures the deviation of the multiplication factor from 1 and therefore enters into the description of non-stationary processes. For the steady-state reactor, reactivity is $\varrho = 0$, and the neutron balance is balanced. Shutdown reactivity stands for the sustained termination of the chain reaction in the reactor core, the long-term holding in the subcritical state) had already fallen below before the start of the experiment – the reactor should have been shut down. In addition, the operating team shut down safety systems. The avoidance of this fault alone would have prevented a catastrophe from occurring.

The explosive power excursion is due to a design flaw in the reactor fast shutdown.

That operating rules were violated is a fact. To what extent they were known to the personnel is questionable. Inexperience and insufficient knowledge were probably the determining factors. The postponement of the test by about half a day contributed significantly to the occurrence of the accident, which made the neutron-physical behaviour of the reactor considerably more complex and unclear.

Similar action steps as in the sorcerer's apprentice can be observed:

- Arrogance coupled with ignorance (disregarding safety regulations),
- Power Rush (expected to be honored on May 1 as heroes),
- Desperate act (continuing even though the test sequence had to be interrupted, thus the possibility of avoiding the catastrophe was not used),
- a sarcophagus will be erected for long-term damage control.

Transformed to the product approach to risk value:

Probability of Damage Reasons for the planning and execution of the deliberately induced attempt are not discernible.

Damage Impact To limit the damage, a sarcophagus with a height of 108 m, larger than the Statue of Liberty in New York, is being erected. The extent of the damage itself is not foreseeable, as late damage is still to be expected.

Risk It was not possible to intervene in the course of the disaster, even the clean-up operations were carried out without adequate protection of the personnel.

The assessments for the sequence of actions and the product approach for the risk value, which were made here for the Chernobyl accident, will be taken up in each of the total of four disasters described and used as a standard of assessment.

The assessments of the action sequences for all four events considered are summarized in Table 1.1.

1.5   Fukushima Daiichi (11 March 2011, Destruction of Several Power Plant Units)

A magnitude 9.0 earthquake occurred off the coast of Honshu at 14:46 local time on 11 March 2011. The focus was located about 130 km south of Sendai and 372 km northeast of Tokyo. The earthquake and the tsunami tidal wave it triggered wreaked havoc in eastern Japan, complicating the damage process and the necessary countermeasures to an unprecedented degree (Mohrbach 2012). Compared to the events in Chernobyl, the boundary conditions were incomparably more difficult, since there a "self-destruction" had taken place at an intact plant and environment.

About 40 min after the quake, the first of several tidal waves reached the power plant. The approximately 30 m high walls of water knocked the entire site short and small. What remained was a rubble landscape like after a bombing raid.

The fuel rods were without cooling, and a hydrogen explosion occurred in the containment and reactor building (Mohrbach 2012).

Two consequences of the Harrisburg incident (Three Miles Island) of 1979 were not taken into account:

No hydrogen decomposition system and no inerting system to prevent explosions and no relief valve for the containment had been retrofitted, although knowledge about their effect on damage reduction was available.

Originally, a 35 m high natural "hill" protected the power plant from tsunami. This was removed by 25 m in 1967 in order to have more favourable traffic routes. As a result, the Fukushima Daiichi power plant was now about 5 m below the tsunami waves of about 30 m recorded from the past (Mohrbach 2012). Motives for the absence of the two safety systems and the terrain removal are to be found in the more cost-effective economy, in the shareholder system. The measures taken by the operating team also reveal a clear lack of safety awareness, which may have been shaped by corresponding management decisions.

Similar action steps as in the sorcerer's apprentice can be observed:

- Arrogance towards the urgent recommendation to retrofit additional safety systems,
- Power rush (consistent implementation of economic interests),
- Act of desperation (measures for the stability of the plant had a higher priority than the protection of the population),
- no calls for help (offers of foreign help are rejected),
- very costly and several decades lasting rehabilitation measures are necessary.

Transformed to the product approach to risk value:

Probability of Damage   Tectonically triggered earthquake with tsunami.

Damage Impact   Efforts were made to limit the extent of personal and economic damage, and in particular the rules for protecting people from radioactivity were consistently followed.

Risk   The disaster was triggered by a natural event. In limiting the extent of the damage, no consistent action was taken in accordance with safety aspects.

The assessments for the sequence of actions and the product approach for the risk value made here for the accident at Fukushima Daiichi will be taken up for each of the four disasters presented and used as a standard of assessment.

The assessments of the action sequences for the total of four events considered are summarised in Table 1.1.

1.6   Explosion of the Deepwater Horizon Oil Rig (20 April 2010)

On 20 April 2010, gas leaked from a BP well in the Gulf of Mexico at 20:52. Eyewitnesses later reported that the gas escaped from the surface of the sea with a strong hissing and bubbling sound, forming a smog-like spray that completely enveloped the drilling platform 30 m above the sea. A few minutes later, a spark, probably caused by a fishing boat below the floating platform, ignited the gas, causing a huge explosion. It severely damaged the rig and burned it out completely, causing it to sink into the sea on April 22, 2010. The explosion killed 11 members of the drilling crew, and caused enormous environmental damage due to large quantities of leaked oil. It was not until the end of August 2010 that the uncontrolled gushing of the oil well was brought under control with the help of a containment vessel (Plank et al. 2012).

How could this disaster have happened?

A major cause of the accident on the BP well was the time pressure the drilling crew was under. The well was 43 days behind schedule on 20 April 2010, the day of the accident, with estimated additional costs by then of around US$30 million. The drilling team therefore sought to complete the well as quickly as possible. It was therefore decided to take an unusual approach. It is interesting to note that this risky approach was approved without hesitation by the US regulatory authority, the Minerals Management Service (MMS), on 15 April 2010. The authority did not even notice that BP's cementing plan did not comply with the legally required minimum cementing distance of 150 m above the reservoir. Obviously, not only BP but also the regulatory authority largely failed in this drilling (Plank et al. 2012).

To make matters worse, it was later discovered that the blowout preventer (well plug at the head of the well), which is designed to protect against uncontrolled escape of oil and gas, had not been maintained and the batteries were dead, thus not completely sealing the well. This negligence is completely incomprehensible, considering that the blowout preventer is the most important safety device in a well. To replace the drilling mud, the drilling crew virtually opened a bubble bottle (Faszination Forschung 2012): gas shot up through the liquid cement (the foam cement had the wrong density, and a foam stabilizer was not used (Plank et al. 2012)), broke through the inadequate pressure protection on the seabed (blowout preventer) and exploded with the drilling rig. Oil and gas were able to flow out of the well unhindered for months and pollute the environment. It was not until August 2010, i.e. four months later, that the oil inflow was stopped with the help of a so-called "static kill" and two weeks later the reservoir was finally and permanently cemented by means of a "bottom kill" (Faszination Forschung 2012).

Similar action steps as in the sorcerer's apprentice can be observed:

- Arrogance coupled with a lack of knowledge ("lack of competencies") (Hopkins 2012), which found its support in the ignorance of the supervisory authority.
- Rush for power, priority implementation of economic interests, the project was consistently implemented (because of the cost and deadline pressure, all proposed measures for damage limitation, which resulted from computer simulations, were thrown to the wind).
- Desperation act, even the installation of a non-functional blowout preventer was accepted.
- Only by means of an expensive auxiliary borehole from the side was the borehole successfully filled and permanently plugged.

Transformed to the product approach to risk value.

Probability of Damage Lack of expertise and deadline and cost pressures can certainly be seen as triggering factors.

Damage Impact Aided by an inexperienced regulatory agency and bad decisions by the drilling team, the largest oil spill to date occurs, costing up to US$90 billion.

Risk Risky actions and actions that contradict the rules of safety are not prevented by the supervisory authority, but tolerated (Plank et al. 2012).

The assessments for the sequence of actions and the product approach for the risk value made here for the disaster on the Deepwater Horizon oil rig will be taken up for each of the four disasters presented and used as a standard of assessment.

1.7   Summary of the Four Events

The assessments of the action sequences for the total of four events considered are summarised in Table 1.1. It shows the common action characteristics of the three disasters described and of the sorcerer's apprentice.

In the case of the events at Fukushima Daiichi and Deepwater Horizon, in addition to the action characteristics shown in Table 1.1, two further factors made a significant contribution to the occurrence of the accident which are not apparent in the action steps in the Sorcerer's Apprentice.

In Fukushima, it was specific manifestations of Japanese culture, ignoring the urgent international recommendation to install necessary safety equipment, and in Deepwater Horizon, that confidence in the actions of the regulator was not justified.

In the case of the Sorcerer's Apprentice, Chernobyl and Deepwater Horizon, human actions can be regarded as having triggered the accident. In the case of Fukushima Daiichi, the disaster triggered a natural event for which humans were inadequately prepared for a possible defence because of the decisions they had made independently in advance.

Human error was also causal in the other catastrophic accidents:

Seveso (10 July 1976, leak of an unknown quantity of highly toxic dioxin or Seveso poison).

Bhopal (03 December 1984, worst chemical disaster and one of the most dominant environmental disasters).

Sandoz (01 November 1986, fire in Schweizerhalle, the Rhine turned red, the landfill still endangers the neighbouring drinking water wells).

Zeebrugge (06 March 1987, sinking of the ro-ro ferry Herald of Free Enterprise).

Überlingen (01 July 2002, collision of two airplanes with a total of 71 people, none of whom survived).

Table 1.1 Compilation of the action sequences of The Sorcerer's Apprentice and the three technical catastrophes discussed above

Event	Characteristics of action		
Sorcerer's Apprentice	Hubris	Ignorance of one's own doubt	Desperate act, with the axe
Chernobyl	Flouting of safety regulations	Power rush, because of expected award	Continue even though the attempt to had to be interrupted
Fukushima Daiichi	Urgently recommended safety upgrades were not taken into account	Consistent implementation of economic interests	Offers of foreign aid were rejected
Deepwater Horizon	Lack of knowledge, supported by an unexperienced supervisory authority	Consistent implementation of economic interests	Even the mitigation measures revealed a lack of knowledge

With the exception of the sorcerer's apprentice, all of these accidents occurred due to unknown weaknesses in the technical system and an unfortunate chain of unforeseeable circumstances that were triggered or amplified by the interaction between humans and the technical system.

The four accident sequences are a confirmation of the "swiss cheese model" developed by the English psychologist James Reason; a pictorial representation (ÄZQ 1990) of latent and active human failures as a contribution to the collapse of complex systems, which describes the concatenation of accident causes (Reason 1994).

Figure 1.1 indirectly shows that the operators pursued an absolutely reasonable objective during the course of the accident and, in doing so, also deliberately switched off installed safety barriers (Chernobyl) or installed non-functional safety elements (Deepwater Horizon) or omitted to install them (Fukushima Daiichi).

If "The Sorcerer's Apprentice" is included, knowledge gaps on the part of the operators or the system designers can be identified in all four accident sequences. The knowledge gaps of the operators (latent and/or active failure) are due to existing or accidental unfavourable system constellations.

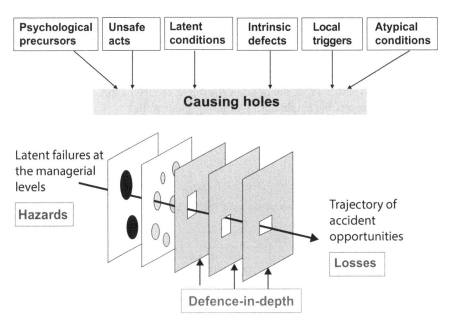

Fig. 1.1 Attempts to capture some of the stochastic features involved in the unlikely coincidence of an unsafe act and a breach of the system's defences. It shows a trajectory of accidental opportunity originating in the higher levels of the system (latent failures at the level of management), traversing the levels of preconditions and unsafe acts, and finally breaking through the three subsequent safety barriers. The figure highlights how unlikely it is that any set of causal factors will find a suitable shoot-through trajectory. (ÄZQ 1990)

However, these accidents also show that the control potential of human action is of eminent importance in minimizing the adverse or devastating consequences of disasters ("accident management").

That is perhaps the most interesting thing about the subject of risk: if one had certain knowledge about decisions, one would not have to decide. You would only have to agree on common goals. In most risk situations, there is a lack of secure knowledge – that is precisely what makes risks.

That is why scientific risk calculations are so important. They make use of the product approach mentioned above. They only provide average values over (theoretically infinitely) long periods of time. When and where a risk manifests itself as the occurrence of damage remains hidden in the fog of probability calculations.

The primary goal of our efforts must therefore be to prevent, or at least limit, the damage identified by risk assessment. Only by applying a benchmark established by social consensus does risk assessment make sense. This reference standard can be very different in different areas. Is it a question of the protection of the individual, the ability of humankind to restore itself, the conservation of biological diversity, contributions to sustainable development, the preservation of biotopes?

In answering these questions, both the response according to the protection goal (e.g. pollutant-free water $\neq$ low-pollution water) and the reference point (such as health, environment, welfare) must be used.

When weighing up the advantages and disadvantages of different possible courses of action in a situation that has been identified as risky, decision-analytical procedures are used. They are used to systematically and explicitly evaluate risks and benefits. These procedures – so-called protection-goal-oriented procedures – are characterized by a pre-considered approach, which is necessary for a comprehensible decision. In this approach, one does not commit oneself to a certain level of risk in advance. Only when the risk value has been determined by analytical steps is it decided, on the basis of the reference standard determined by social consensus, whether the risks associated with the project are acceptable or unacceptable.

This decision is based on basic assumptions, "mental patterns". "Mental patterns" describe how things are thought about, what meaning we attach to things, how we behave and how we see the world around us. "Mental patterns" imply the reduction of complexity by limiting it to easily manageable fields and resorting to familiar facts by means of analogy. "Mental patterns" are not subject to discussion. We use them unconsciously in all areas of life, and they occur so quickly that the conscious mind does not even register this perception. "Mental patterns" control our social behavior, especially our empathy. More than a hundred years ago, Charles Darwin already suspected that empathy, as a precursor to active compassion, is an important survival tool in nature's repertoire. Empathy is a bonding agent of social cohesion, and humans are the social animal par excellence.

In physics one would say it is resonance, the so-called resonating. Psychology has coined the term social intelligence for the synchronization of these processes. The synchronization of interpersonal behavior is achieved through the order of the contents of life – culture. Culture has an effect in various factual areas. In the area of dealing with risk-generated technology, one speaks of safety culture. Risk-oriented knowledge improvement through safety culture serves to reduce the uncertainties remaining in the system and thus to reduce the inevitable risk. Safety culture can be lived, for example, through traffic education.

In connection with the four individual events, the question arises: How do we humans deal with our responsibility?

The notion of personal responsibility is strongly ingrained in Western culture, but it should be borne in mind that people involved in serious accidents were neither unreliable nor reckless, even if they may have been ignorant of the consequences of their actions. We must also not fall prey to the fundamental attribution error of blaming people and ignoring situational factors.

Unintended consequences of technical systems, of new inventions, are an inevitable side effect. Each new generation of technical devices floods society before we really know what changes it will lead to. Thus, every innovation is also always an ongoing social experiment. A one-sided view of societal goals, as is currently the case in the financial industry, ignores the emotional bonds of people that foster our abilities to feel good and do our best.

Just as the technology depicted shows a differentiation of evolution, the sorcerer's apprentice is a milestone of spiritual culture and also serves as a value ordering the human community.

What did you miss?

An answer to the question, what is the role of the sorcerer and is there, if any, direction in the events of Chernobyl, Fukushima Daiichi and Deepwater Horizon or are these events a matter of handling responsibility or reliability?

An answer to this question is attempted in Chap. 2.

References

ÄZQ (1990). www.patientensicherheit-online.de (Übersetzt aus: Reason J (1990) Human error. Cambridge University Press, Cambridge). Accessed: 1. Dez. 2018

Faszination Forschung (2012) Das Wissenschaftsmagazin der Technischen Universität München, Ausgabe 11, S 52–65

Hauptmanns U, Herttrich M, Werner W (1987) Technische Risiken. Springer, Berlin (Geleitwort von Prof. Dr. Klaus Töpfer)

Hopkins A (2012) Disastrous decisions: the human and organisational causes of the Gulf of Mexico Blowout. Oxford University Press, Oxford

Mohrbach L (2012) Seebeben und Tsunami in Japan am 11. März 2011. VGB PowerTech, Essen

Plank J, Bülichen D, Tiemeyer C (2012) Der Unfall auf der Ölbohrung von BP – Welche Rolle spielte die Zementierung? TUM, Lehrstuhl für Bauchemie, Garching, Deutschland

Reason J (1994) Menschliches Versagen. Spektrum, Heidelberg

Scoerer's Apprentice (2018). http://www.unix-ag.uni-kl.de/~conrad/lyrics/zauber.html. Accessed: 2. Dez. 2018

Cause-Effect Structure and Intentional Structure

2

2.1 Introduction to the Question of Directing

The previous chapter concluded with the question; who actually directed the four events depicted?

This question has occupied mankind since the beginning.

Dschung Dsi wrote 2000 years ago (Prigogine and Stengers 1990): "The circulation of the heavens, the persistence of the earth, the way the sun and moon follow each other in their orbits: Who is it that rules them? Who is it that binds them together? Who is it that keeps far without toil and all this going? Some think it is a driving force that causes them to be unable to do otherwise ...".

The question of cause is aimed at the basic model of the world, which according to Thomas Bartelborth consists of the following four components (Bartelborth 2007):

(a) Items,
(b) their (intrinsic) properties,
(c) of the spatiotemporal structure,
(d) the cause-effect structure.

We will try to answer Dschung Dsi's question via the cause-effect structure and the spatiotemporal structure, both natural science findings, and via the intentional structure, representative of the social science field.

The four individual events described in Chap. 1 each have different meanings for different people.

For natural scientists, they offer insightful approaches to the underlying, possibly still hidden control processes, which can be opened up through the cause-effect structure and the spatiotemporal structure.

V. Hoensch, *The Chernobyl, Fukushima Daiichi and Deepwater Horizon
Disasters from a Natural Science and Humanities Perspective*,
https://doi.org/10.1007/978-3-662-65319-7_2

For the social scientist, the same events are the greatest threat posed to humanity by the use of risky technology. They attempt to use intentional structure to point out errors and mistakes in human action. In doing so, they too acknowledge that the development of human society has come about through the cognitive use of the forces of nature and that social interaction is essentially determined by the development and use of the tools associated with it, Bubb (2007), private communication.

This chapter argues that with the cause-effect structure and the intentional structure, i.e. deductively (deriving the individual case from the general), the four individual events can be connected, which at first glance look as if they stand unrelated to each other (Hume: "the cement of the universe") (Mackie 1975) and thus the question of direction and the associated question of reliability can be answered.

Chapter 3 will examine whether the spatiotemporal structure can also contribute to an answer to the question posed by Dschung Dsi and possibly provide a temporal order and thus "reconcile" the differences between the two different approaches.

But first we describe our understanding of the cause-effect structure and the intentional structure. After that, we will try to establish heuristics for both structures.

2.2   Fundamentals of the Cause-Effect Structure

From the spectrum of scientific approaches, as already stated, we use the cause-effect structure in this chapter.

In Chap. 3 we keep the scientific approach but relate it to the spatiotemporal structure.

The problem for describing the cause-effect structure lies in the fact that neither in philosophy nor in the natural sciences has there been a uniform, sufficiently strong concept of cause to date, according to which causal processes are at the same time necessary (deterministic – reversible) and temporally directed (probabilistic – irreversible).

"The law of causation states that every change has a cause, that every event is linked to a sum of circumstances, in the absence of which it cannot occur and in the presence of which it occurs with necessity" (Kausalgesetze; Meyers Lexikon, 6th volume, 1927). "Determinism, philosophical doctrine according to which the whole of nature, including man and his actions, is subject to the law of causation, which applies without exception; in a narrower sense, any philosophy or religion which maintains that man does not possess freedom of will is called deterministic (Determinismus; Meyer's Encyclopedia, Vol. 3, 1925)".

We will discuss the aspects addressed by the two definitions, such as free will, in more detail. For the time being, we will concentrate on the physical aspects.

From physical signal transmission to neuronal mechanisms, only causal processes are known, which are alternately one or the other, but not both at the same time (Falkenburg 2012). According to the causal principle, every effect has a natural cause. The causal principle underlies all scientific research. It has been typical for physics since Galileo and Newton, for chemistry since the end of the search for the philosopher's stone, and for biology since Darwin. "Nature" is now called physis, and natural science restricts its explanatory grounds to the physical world. Speaking of the philosopher's stone: The causal principle is not a philosopher's stone, but neither is it a myth; it is a general hypothesis subsumed under the universal principle of determination; in the domain belonging to it, however, it possesses an approximate (angenäherte) validity (Bunge 1959).

The debate about cause and effect and their mutual linkage, the causal principle, can be traced back to Aristotle, although he already distinguished four types of causes. This four-cause theory is apparently primarily oriented towards human action – towards what we bring about through technology ("techne"), not towards what happens in nature (physis) by itself (Falkenburg 2012).

We encounter the differences between technology and nature in classical thermodynamics with the irreversibility of energy transformation in all physical processes and in the biological sciences with Darwin's theories on the origin of species based on cumulative changes in the living world.

Based on this, very different currents in science developed until today to formulate the cause-effect structure. The cause-effect structure is a relationship between events – not between properties or states, let alone between ideas (Bunge 1959). Every effect is somehow produced by its causes. In other words, causation is a kind of event generation, or again, put differently, of energy transfer. The causal generation of events is law-like, and the laws of causation can be represented in differential equations. Events, in particular, can modify probabilities: The world is not strictly causal, although it is determinate: Not all interrelated events are causally related, and not all regularities are causal in nature. Causality, the perception of cause and effect, seems to be a structure of the process of human understanding rather than a description of reality. Thus, for Mario Bunge, the cause-effect structure is only a variant of determinism (Bunge 1959).

To date, there are at least four causal terms, we quote Brigitte Falkenburg (2012):

1. The traditional causal principle; a given effect proceeds from the sought cause.
2. The deterministic events according to a strict law e.g. Newton's law of gravitation, Maxwell's electrodynamics, Boltzmann's impact approach or the development of the quantum mechanical wave function Ψ according to the Schrödinger equation.

3. Irreversible processes that are only probabilistically determined. These include thermodynamic processes associated with an increase in entropy, their statistical explanation according to kinetic theory and Boltzmann's H-function, as well as the measurement process and the probabilistic interpretation of quantum mechanics. Here, what happens in a particular case is random. It reflects the temporal order of cause and effect, but not their lawful connection.
4. Einstein's causality (special relativity).

Therefore, there are also doubts about the universal validity of the cause-effect structure: Given the universality of randomness and spontaneity, there are limitations. We cannot and do not want to go into such limitations here; examples are the chaos theory, the probabilistic approaches, the state-space approaches and the TCP-theorem (also CPT-theorem, stands for "charge, parity, time" ≡ charge, parity, time and is a fundamental physical law).

Nevertheless, we use the categories of cause and effect here, and likewise the causal and non-causal relationships between cause and effect. We can often estimate mathematically the probability with which a particular event will occur, but we must not confuse probabilities with causalities.

Almost all phenomena observed in nature behave irreversibly. The amazing thing about this is the fact that the underlying laws of nature do not distinguish any direction in time – they would allow the shards of a glass that fell on the floor to reassemble. But such a process has never been observed.

At this point, a digression is allowed.

If one examines this problem more closely, one finds that of the processes represented as irreversible by the Second Main Theorem of Thermodynamics, none is really irreversible, but merely that the probability of reversible processes is so extremely low that they do not in fact occur. It explains why sugar dissolves in coffee, but this process never occurs in reverse. We cannot "unstir" the sugar in the coffee cup once it has dissolved. It is perfectly consistent with the laws of physics for a dissolved piece of sugar to "unstir" and turn back into a cube when stirred. The probability of such a thing occurring is so small that it can usually be neglected. The Second Main Theorem of Thermodynamics also explains that all things wear out and cool down, that they slacken, are subject to a process of aging and decay. Only through constant energetic expenditure can the system be kept in the desired state and the decay of information as well as the diffusion of responsibility – both terms we will deal with later – be counteracted.

This digression needs to be extended with regard to the concept of the Second Main Theorem of Thermodynamics and the entropy that is directly related to it:

Thermodynamics is a branch of physics which, starting from the study of heat phenomena (in the sense of thermodynamics), investigates all processes associated with the conversion of energy of various kinds and their application. The main laws of thermodynamics are formulated as postulates, which are, however, supported by all experimental experience. The Second Main Theorem of Thermodynamics (entropy theorem) gives the direction of thermodynamic changes of state. Entropy is a fundamental thermodynamic quantity of state with the SI unit of joules per kelvin, i.e. $J \cdot K^{-1}$. The concept of entropy is often colloquially described as being a "measure of disorder". However, disorder is not a physical measure. It is better to think of entropy as a "measure of ignorance of atomic state", although ignorance is not a physically defined concept either. Entropy is thus essentially a statistically defined quantity, and so can be usefully employed in many contexts. Notwithstanding this, definitions may vary across disciplines.

Back to the phenomena observed in nature.

Nature seems to have an inherent arrow of time that is not known to the fundamental laws of physics. These fundamental laws, as we have noted, take the form of differential equations. Their time-reversal invariance means that for every solution of the equations there is a time-reversed version that is also a solution.

Natural processes are either reversible and deterministic or irreversible and indeterministic, but never both at the same time. According to the usual view of the laws of nature, the deterministic, reversible laws of physics can only be applied to irreversible non-strictly determined processes without contradiction by means of tricks: by choosing the correct initial conditions, restricting them to "physical" solutions, and by means of the probability calculus, which is knitted in such a way that it respects the arrow of time (Falkenburg 2012).

Thermodynamic processes – that is, all heat and diffusion processes and all radiation phenomena – are irreversible. No sugar coffee segregates by itself into sugar and coffee and extracts heat from the air in the process. This would be a process described by the "unphysical" solutions of Maxwell's equations mentioned above (Falkenburg 2012). In other words, processes like the one with the broken glass or with the sugar in the coffee are not completely ruled out, but they are so extremely unlikely that they practically do not occur and we rule them out.

Probability says something objectively about possibilities becoming factual and subjectively something about the occurrence of expectations (Falkenburg 2012).

What arrows of time are observed in nature?

The growth of disorder or entropy with time is a favorite example of the phenomenon we call the arrow of time, of something that distinguishes the past from the future by giving direction to time. From this point of view, we appropriate the direction in which world events evolve, it defines the meaning we give to our lives.

Our time consciousness plays a key role for the integrative performances of consciousness and their explanation from neuronal foundations, because what we experience as the present is, after all, identical with our contents of consciousness. What we focus our attention on is present; what we store in our memory is past; what our plans and intentions are directed towards is future (Falkenburg 2012).

A variety of different arrows of time are discussed in science. According to Stephen W. Hawking (1994) alone, there are at least three different arrows of time: (1) the thermodynamic arrow of time, the direction of time in which disorder or probability (entropy) increases; (2) the psychological arrow of time, the direction of time in which we feel time is progressing, the direction in which we remember the past but not the future; (3) the cosmological arrow of time, the direction of time in which the universe is expanding and not contracting. Our consideration in this chapter will be limited to the thermodynamic and psychological arrows of time. The cosmological arrow of time will be the subject of Chap. 3.

The most common explanation of the direction of time is based on the thermodynamic arrow of time, on the Second Main Theorem of Thermodynamics and its justification by Boltzmann's H-theorem by means of classical statistical mechanics.

The explanations of the Second Main Theorem of Thermodynamics are to be deepened.

Thermodynamics, as a branch of physics, is described by three main theorems. The Second Main Theorem of Thermodynamics specifies the direction of thermodynamic changes of state. There are different formulations of the Second Main Theorem of Thermodynamics. Let us understand the Second Main Theorem of Thermodynamics with simple considerations. The separated components yolk, albumen and eggshell form the egg, a system of low entropy but high order (separation of components). There are many ways the egg could fall, break, or perhaps survive the fall. The possibility of surviving the fall unscathed is unlikely, but not impossible. This brings probability into play. The egg analogy can be applied to any natural system. From our everyday observation, we know that some processes only go this way and never the other way around.

Boltzmann was the first to recognize that one could see in the irreversible (nonreversible) increase of entropy the expression of a growing molecular disorder. Boltzmann's idea was thus to relate the entropy S to the number of possibilities: Entropy characterizes each macroscopic state by the number of ways to reach that state P. Boltzmann established the following equation:

$$S = k \cdot \lg P$$

The proportionality factor k, is the Boltzmann constant (a universal constant). The logarithmic expression lg indicates that entropy is an additive quantity.

Thus, according to the Second Main Theorem of Thermodynamics, entropy increases. Thus we succeed in defining the thermodynamic arrow of time: The future lies in the direction of time in which entropy increases. Thus we are familiar with experiences according to the egg analogy, although they could theoretically also run quite differently. But because such courses are so improbable that they most probably do not occur in the entire history of mankind, of the universe, we exclude them. The Second Main Theorem of Thermodynamics is an example of a general law of nature that allows exceptions to a "law". After this digression into understanding the Second Main Theorem of Thermodynamics, let us consider its possible derivations.

Without the Second Main Theorem of Thermodynamics, neither the metabolism of living beings, nor the functioning of technical devices, nor the firing of neurons, nor the mechanism of the mental clock in our brain can be understood (Falkenburg 2012).

The explanation of the thermodynamic arrow of time by metaphysics (what lies behind physics) in some respect harkens back to the difference of earlier and later that we ultimately state with our mental clock (Falkenburg 2012).

Objective, physical time lies in the directed, measurable processes of thermodynamics, quantum physics, and the cosmological evolution of a universe with a uniform world age (Falkenburg 2012).

The psychological arrow of time, based on subjective, mental, and experiential time, lies in the experience of earlier and later events, of simultaneously and consecutively, in the memory of the past, in the experience of the now, and in the anticipation of futures (Falkenburg 2012).

Stephen W. Hawking's (1994) reasoning says: Only if the thermodynamic and the psychological arrow of time (as mentioned above, we leave the cosmological one out of consideration in this chapter) point in the same direction, the conditions are suitable for the development of intelligent living beings. A pronounced thermodynamic arrow of time is a necessary precondition of intelligent life. To live, humans must take in food, which is energy in ordered form, and convert it into heat, energy in disordered form (see Sect. 2.10). Our subjective sense of the direction of time, the psychological arrow of time, is determined in the brain by the thermodynamic arrow of time, so that the two always point in the same direction (Hawking 1994).

The study of the problems connected with our experience of time is complicated by the fact that no physical conditions are the direct stimuli for mental reactions. Man has no sense of time resembling the sense of sight or hearing, and accordingly no organ that can be assigned to the subjective experience of time. Our consciousness and the experience of the present are closely related, but both remain mysterious.

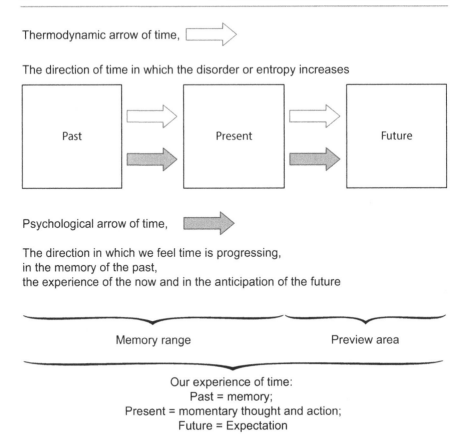

Thermodynamic arrow of time,

The direction of time in which the disorder or entropy increases

Psychological arrow of time,

The direction in which we feel time is progressing,
in the memory of the past,
the experience of the now and in the anticipation of the future

Memory range   Preview area

Our experience of time:
Past = memory;
Present = momentary thought and action;
Future = Expectation

Fig. 2.1 Illustration of thermodynamic and psychological arrow of time

Our experience of time teaches us that time passes, that everything present will soon be past, and that we can make plans directed toward future actions and events. Intentionality is nothing more than the direction of consciousness toward future things that are to become present. If the direction of our experience of time cannot be explained physically, then with it the whole of intentionality remains non-derivable, non-reducible (irreducible), but nevertheless intentional structures remain indispensable for the explanation of the four individual events, as we shall see.

The experienced time is subjective time, is consciousness of the present, the past, the future. At the same time, we know that only the moment is real, the past is already over and the future has not yet occurred. With measured time we keep to the clock in the course of the day and to the calendar in the course of the year in order to coordinate ourselves in everyday life (Falkenburg 2012; see Fig. 2.1).

The arrow of time in physics is objective time. The time structure is fundamental for our subjective experience, and it underlies our actions. All this against the background that the physical arrow of time remains mysterious to this day! (Falkenburg 2012).

The direction of time, that is, what we would like to have seen explained, defies any physical explanation (Falkenburg 2012).

Only the cosmological principle allows to define a universal, uniform, cosmic time. It must be taken into account that the cosmological principle can hardly be tested empirically. We will discuss this principle in more detail in Chap. 3.

The explanation of the Second Main Theorem of Thermodynamics by the kinetic theory already presupposes the difference of earlier or later, i.e. the time direction. The kinetic theory of gases is based on the free and unregulated motion of molecules or atoms in the gaseous state and derives from this their equilibrium properties such as temperature. According to this theory, the temperature of a gas corresponds to the average kinetic energy of the molecules, and the entropy corresponds to the probability of the distribution of the molecular states. This explanation of the Second Main Theorem of Thermodynamics does not help us to the sought explanation of the objective order of time, but this order of time conversely enters into the explanation of the Second Main Theorem of Thermodynamics by the kinetic theory. And this is a problem of principle. No deterministic theory describing reversible processes can explain irreversible processes without specifying their direction at least by suitable initial conditions (Falkenburg 2012).

In short: The Second Main Theorem of Thermodynamics, the increase of entropy, and thus the thermodynamic arrow of time, can be read in sociological terms as a law of increasingly rational action orientation (Müller 1996).

For the cognitive science view, we refer to the intentional structure.

After working out the basics of the intentional structure, we will try to present heuristics for the cause-effect structure and for the intentional structure. In later steps we will use both heuristics to scrutinize the four single events and thus be able to answer Dschung Dsi's question. Heuristics are important in recent philosophy of science as a criterion for evaluating theories and entire scientific programs.

2.3   Fundamentals of the Intentional Structure

The previously mentioned Aristotle already described the action as an intentional process. He distinguished the intention of action from the anticipated goals of action and the circumstances of action.

Fundamental to the description of intentional processes is the identification of goals, their hierarchical organization and the necessary feedback loops.

With these definitions, we draw on the differences between natural science and cognitive science approaches.

Following the research tradition of natural science approaches, we have presented the basics of the natural law cause-effect structure. This tradition also forms the basis for much of our knowledge about human cognition, i.e. the perception, recognition and processing of information.

Before we take up the cognitive science approach, which is concerned with broad scaffolding rather than clearly laid out data-oriented models such as those provided by the cause-and-effect structure, it is necessary to distinguish between actions without prior intention, that is, spontaneous and skill-based actions, and intentional actions.

(Skill-based is also translated as "ability-based"; we do not agree with this translation. We will discuss the differences between skill and ability in Chap. 3).

Searle (2006) made this important distinction between actions without prior intention, that is, spontaneous and skill-based actions, and intentional actions. According to this, the former are highly practiced unconscious sequences of actions. Intentional actions, on the other hand, are conscious actions shaped by the will. This distinction is also important for the question of reliability, which is the focus of our analyses.

Accordingly, reliability is closely linked to the intention to act, the intention. On the other hand, actions are quite conceivable which contradict reliability, but which lie outside our consideration.

In this context, the distinction between intentions to act, i.e. reasons and purposes, as opposed to causes, as highlighted with the application of the cause-effect structure, is again emphasized. It is necessary to distinguish neatly between intentions to act, intentions and causes. Intentions to act are reasons, not causes (Falkenburg 2012). The search for the physical causes of our action processes, which the cause-effect structure demands, is not compatible with the specification of our intentions to act because, according to Wingert (2006), the concepts of purposes, reasons, goals, and intentions are anthropocentric concepts that have no place in scientific explanations. Purposes are something subjective. They are based on human intentions – on motives, desires, plans, intentions and orders, by whomever ("par ordre du mufti").

At this point, a digression on the phrase "par ordre du mufti" is allowed.

The expression "par ordre du mufti" is colloquially understood as an action by decree, by order of a superior authority, by foreign command or by necessity. Practically analogously, the word "ukaz" (Russian "ukaz" to "ukazat", to command) is understood as an order, command, or decree of the tsar, now the president of the Russian people's representative body, usually ironically. Both the idiom and the word are commonly meant to express that a decision is made "from above" without hearing those affected or enforcing it explicitly against their vote. We will discuss the aspect of stakeholder involvement or enforcement against their vote in detail in Chap. 4. At this point we return to the intentions for action.

Activities are realised in actions (Hacker and Richter 2006). Action refers to a temporally self-contained unit of activity that is directed towards a goal and structured in terms of content and time. Action control is the result of the complex interaction of genetic predispositions, learning experiences, currently processed stimulus information and the momentary motivational state of the individual (Franken 2007).

Cognitive psychology describes the process of perception, recognition and processing as information processing, in which the brain (consciousness) is also involved in addition to the sense organs. Perception is not only understood as the reception of external stimuli, but also the subjective construction of one's own world view based on the sensory impressions from the environment (Franken 2007). Cognitions are tools with which people find their way in the world.

In the current information processing models, the cognitive processes are described in different ways. What these models have in common is the division into different phases that form a control loop, which we summarize in this way:

Information reception (stimulus-stimulus level [environment]), sensory-perceptual level (perception), information processing (cognitive level, thinking – storing – remembering), information conversion (motor-effector level [acting]), effector level; this includes the organs that carry out reactions and information coding on the basis of information evaluation.

We rely largely on Jens Rasmussen (1986) because his account of cognitive control mechanisms is error-oriented. Rasmussen's model targets errors committed in the supervision of industrial plants, especially in accident-prone situations such as those that occurred in our selected individual cases.

The actions necessary to perform a task are the result of the mental information processing level. They are influenced by an overlying social level and an underlying psychological level. Subjective values are formed on the social level and thus action goals are given for mental information processing. The psychological level comprises the actual mental processes that provide the tools for human information processing. The mental processes are cognitive and sensorimotor, besides affective (i.e. emotional) in nature. Mental processes have different cognitive levels. If they do not run consciously, then we speak of automatic or automated processes with a very low cognitive level. If the mental processes are consciously guided, then we speak of cognitive processes in the narrower sense or conscious cognitive processes. These are also called controlled processes or cognitive processes as well as mental processes. This is represented by the model of Rasmussen (1986), with which he treats the processes of mental information processing in a differentiated way with respect to the degree of cognitive engagement of the human being. With the differentiation of skill-, rule- and knowledge-based action, which is regarded as "standard", the preoccupation with human reliability also succeeds.

Rasmussen distinguishes three levels of cognitive behavior or ability, namely:

At the skill-based level, human performance is determined by stored patterns of pre-programmed instructions represented as analogous structures in a spatiotemporal domain – of which more detail is given later.

The rule-based level comes into play when addressing familiar problems where the rule of the type if … (state), then … (diagnosis) or if … (state), then … (action) is determined by conscious action in familiar situations.

The knowledge-based level comes into play in novel situations where actions must be currently planned and executed using conscious analytical processes and stored or to-be-acquired knowledge.

A key criterion for the distinction, which builds on Rasmussen's levels of execution, is whether or not an individual was engaged in problem solving at the time an error occurs. Behavior at the skill-based level represents activities that occur as consistent, automated, and preprogrammed patterns after an intention has been set without conscious control (Reason 1994). Errors at this level result because the knowledge to respond to change is not accessed at the right moment.

Rule-based and knowledge-based executions occur after the individual becomes aware of a problem, i.e., the unanticipated occurrence of an event that requires deviation from non-conscious, skill-based action. In this sense, problems at the skill-based level precede problem solving at the rule- and knowledge-based level. The transitions between skill-based, rule-based, and knowledge-based behavior are slippery and blurred. Skill-based behavior can also occur at the knowledge-based level. The consequence of this is that ingrained accuracy that exists at the skill-based level is then lost. Thus, a defining condition for both rule-based and knowledge-based errors is awareness of the existence of a problem (Reason 1994). Rule-based errors lack the knowledge of when deviations from routines exist and in what forms they occur. At the knowledge-based level, errors result from changes in the work process for which one is not prepared and which one could not anticipate.

Rasmussen's three cognitive levels, which can also occur simultaneously, are associated with a decrease in familiarity with the environment or with the task (Rasmussen 1986).

As expertise increases, the human control center moves from the knowledge-based to the skill-based level; however, all three levels can coexist at any time.

James Reason dedicated his book "Human Failure" (Reason 1994) to Jens Rasmussen, from which we quote:

> Rasmussen identified eight stages of decision making (or problem solving): activation, observation, identification, interpretation, evaluation, goal selection, procedure selection, and execution. While other decision theorists present these or comparable stages in a linear sequence, Rasmussen's major contribution was to capture the shortcuts people take in making decisions in real-life situations. Rather than a straight-line sequence of stages, Rasmussen's model is analogous to a stepladder, with skill-based stages of activation and execution at the bottom of both sides and knowledge-based stages of interpretation and evaluation at the top. In between are the rule-based stages (observation, identification, goal selection, and procedure selection) on either side. Shortcuts can be chosen between these different stages, usually in the form of very efficient but situation-specific stereotypic responses, where observation of the system state automatically leads to the selection of a procedure that provides relief, without the slow and tedious intervention of the knowledge-based processes. The "stepladder" model allows for associative jumps between all levels of decision making and thus freedom of action. (Reason 1994)

Let us graphically summarize the three levels of execution and the "stepladder" model.

In order for the associative leaps to become effective for regulating action, the agent must voluntarily orient his behavior accordingly. There is no recognized definition for the term voluntariness. Colloquially, one understands something different by free will than in legal or psychological usage. In philosophy, the term is not uniformly defined.

Man feels himself to be free, he is subjectively free – this is true on the one hand because he cannot estimate the conditions and causes of his will, but on the other hand also because he cannot precisely predict the consequences of his will and his actions, but can only use statistical empirical values. This does not change the fact that objectively every decision of will is determined by causes. We are our brain, and our brain chooses between alternatives – and it does, on the basis of fixed parameters, quasi as a computational process. This is the objective process that we call freedom from the human perspective.

In a multidisciplinary sense, freedom of will includes the subjectively perceived human ability to make a conscious decision when faced with various choices.

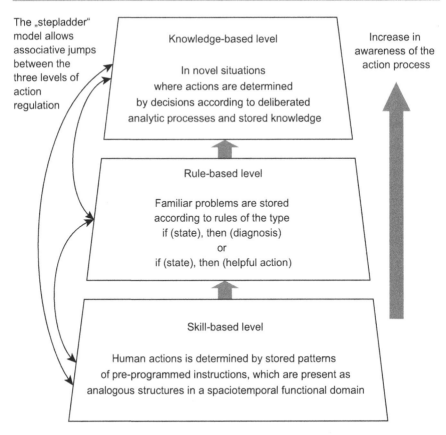

The "stepladder" model allows associative jumps between the three levels of action regulation

Knowledge-based level

In novel situations
where actions are determined
by decisions according to deliberated
analytic processes and stored knowledge

Increase in awareness of the action process

Rule-based level

Familiar problems are stored
according to rules of the type
if (state), then (diagnosis)
or
if (state), then (helpful action)

Skill-based level

Human actions is determined by stored patterns
of pre-programmed instructions, which are present as
analogous structures in a spaciotemporal functional domain

Fig. 2.2 Rasmussen's Skill-Rule-Knowledge model specifies people's regulation of action via three hierarchically arranged levels of execution of human behaviour. A key development is that Rasmussen considers shortcuts that people take in making decisions in real-life situations ("stepladder" model), allowing for freedom of action

We discuss Benjamin Libet's epoch-making experiments on free will in detail in Chap. 3 (Fig. 2.2).

Voluntariness is ultimately based on our drive to fathom: "That I may know what holds the world together at its innermost, show all agency and seed." in Chap. 3. "What the world really is, we don't know anyway" (Lesch 2016). What drives us are the gaps, the gaps in knowledge. Such gaps in knowledge do not exist in determinate behavior because here our actions are determined by the cause-effect structure and thus nothing can be decided within this structure. Something we do not know is what leads us to voluntariness. Voluntariness creates the possibility for interpretation and thus doing things that are not clearly determined. Society judges with its view, scientific or another, whether the possibility for interpretation has been used correctly.

What tensions social judgement can be exposed to in this process by using the separable possibility of interpretation, we would like to illustrate by the dispute between Newton and Goethe, which the latter conducted fiercely. Newton said that sunlight consists of rays of different refractive power and that the white of sunlight is composed of spectral colours. According to Goethe, light is a "unity," and color is a phenomenon of different quality. From today's point of view, Goethe's and Newton's theories of colour arise from two irreconcilable points of view, the subjective one, which man cannot escape (Goethe) and the objective rational physical one (Newton). In connection with this point of view a short review to the arrow of time; Goethe's point of view stands for the expectation, Newton's, however, for the preview into the future. To understand, we need to juxtapose both views and see them in historical context. Science stands for a specific set of methods of finding out about something that admits of systematic investigation.

In short, there are only the facts that we have found out in a specific way and thus know (Searle 2006).

The existing knowledge gaps are related to the probabilistic character of quantum processes and the unsolved measurement problem. We will discuss the probabilistic character of quantum processes in more detail. The unsolved measurement problem is the subject of quantum physics, which we will not discuss here. With this understanding of science and the principle of intentional, voluntary action, we bump into these gaps in knowledge. Gaps in knowledge require a spontaneous order that sets general rules for individual action; that do not prescribe a general goal, but merely set mutually recognizable limits to the pursuit of specific goals. Thus, rules of the road do not prescribe a destination for road users. Their coordinating function is merely to create, by means of appropriate precepts, conditions under which all can smoothly achieve their individual goals.

Thus be it said otherwise:

First, scientific knowledge remains a patchwork of theories and laws until the gaps in knowledge are largely closed, valid in their respective fields but at best loosely connected to theories and laws in other fields.

Second, there is also no guarantee that all scientific knowledge can be assembled into a single picture. Certainly, there are many scientifically based pictures of science (Searle 2006).

With voluntariness, barriers – which are so beautifully illustrated in the well-known "Swiss cheese model", see Fig. 1.1 of Chap. 1 – can of course also be overcome. But these barriers, too, are of course subject to the law of increasing entropy. This means that they fail by chance – probabilistic influencing factors come into play here – or that external states are added by chance for which no barriers were intended. It is "merely" that gaps in the barriers (the holes in the "Swiss cheese model") have not been detected or have not been detected in time (occurrence of probabilistic influencing factors) or that existing gaps in knowledge have been encountered by chance. The non-recognition of gaps naturally also includes the fact that such barriers are skipped in order to achieve some kind of advantage, since in general all (!) barriers serving security are associated with inconvenience and additional effort. What is seen as an advantage from the point of view of the agent can of course also be very burdensome from the point of view of a larger community/society, as we will make clear with the three incidents discussed.

The randomness of overcoming the barriers by which the desired orderly flow of a technical system is to be ensured illustrates the need to address the problem of "Who is directing this?" and also to include the aspect of reliability. Direction is the guidance and supervision given by an experienced person. There are various synonyms for the word direction in different fields. We have chosen "par ordre de mufti". Reliability or dependability generally refers to the dependability of people, apparatus and materials that can be trusted. As a human trait, it is considered a virtue of character. In technology, reliability is a characteristic of technical products.

Based on the basic structures for the cause-effect structure, based on the thermodynamic arrow of time, and the intentional action, based on the psychological arrow of time, we will then work out specific heuristics.

To summarize: the question of direction can only be applied to intentional actions, that is, to actions at the rule-based or knowledge-based level. It has no meaning for skill-based or unconscious actions. Regarding direction: Man is the actor in natural events, he has to play his role, even if he occasionally cannot or does not want to cope with it, he has to play his role if he does not want to fail.

But who is the director? Who wrote the script? And who says when which actor has to step where on the stage of the natural event?

> "The directors, that is, the ones who tell where things are going, these are the forces (four forces; the electromagnetic force, the strong and weak nuclear force, and gravity; insertion by author). At least according to scientific knowledge, the forces are those that cause particles to form certain structures." (Lesch 2016).

Rupert Sheldracke (2015) has also addressed the question of direction. He writes: "It is true that all material things are made of quantum particles, that there are flows of energy associated with all physical processes, and that all physical events take place within the framework of spatiotemporal given by the universal gravitational field."The four elementary forces direct. Man becomes a "director" by the fact that causality is assigned to his actions. The laws of nature in their entirety write the script. Entropy and the thermodynamic arrow of time derived from it are the "script-writers". Through our intentions we tell which actor has to step on the stage of natural events when and where.

Recall Francis Bacon's "Golden Rule," set down in his 1620 work Novum Organon: "Natura enim non nisi parendo vincitur (Nature can only be defeated by obedience)."

The script is written by the laws of nature, this can be clearly experienced with the evolutionary development on our earth.

We will explore this scientific approach to directing and screenwriting in more depth later in this paper.

2.4   Establishing a Heuristic for the Cause-Effect Structure

There are different views of heuristics in the various fields of application, and we have deliberately refrained from formulating them here. For the pragmatic approach we have chosen, the focus is on creative or problem-solving activity; it can thus be regarded as an optimization procedure.

The elements of the cause-effect structure, cause, effect, are to be regarded as single events, of which one causes the other however. We know a priori that the cause happens earlier than the effect – the objective time order (asymmetrical order). Thus, we are able to temporally line up the processes we perceive and experience (Falkenburg 2012). Causality is the connection of cause and effect and lies in general laws of nature.

It should be noted that there is no agreement among scientific disciplines on exactly how to understand the link between cause and effect.

A relatively high degree of unity is found in physics and chemistry. It becomes weaker with biology and medicine and quite weak with philosophy. You can see from this that the more animate nature comes into play, the more difficult unity becomes.

We can now refer to Kant's epistemology, which establishes a close connection between the objective order of time, i.e.: the thermodynamic arrow of time, and a connecting principle – the causal principle. Kant noticed that causal processes are temporally directed or irreversible; the effect follows the cause and never vice versa.

Even physics cannot clearly and unambiguously specify the concept of causality. It offers several causal concepts.

According to Newton's mechanics, Maxwell's electrodynamics and Einstein's relativistic physics, the relationship between cause and effect is deterministic but reversible, i.e. time-symmetric. According to thermodynamics and according to any

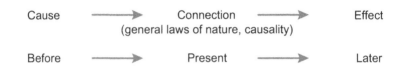

Fig. 2.3 Only the arrow of time of thermodynamics is known to physics as it stands today

quantum theory, this relation is irreversible for single events, thus time asymmetric, but not deterministic. This is due to the fact that macroscopically observable single events are based on the interaction of an extremely high number of microscopic elementary particles.

If one examines the problem more closely, one finds that none of the processes observable as irreversible is really irreversible, but, as already explained, that only the probability of this is extraordinarily low.

The aspect of probability will be dealt with in more detail and for each individual event specifically for the probabilistic influencing factors in a later section.

Our heuristic applied to the cause-effect structure is thus as follows, cf. Fig. 2.3.

2.5   Establishing a Heuristic for the Intentional Structure

For setting up a heuristic for intentional structure, we borrow heavily from Hacker and Richter (2006).

We quote:

> Activities or actions are components of behaviour in whose mental regulation all manifestations of the mental are involved. The concept of activity can be understood as an umbrella term, while actions denote (through goals) delimited units of activities.

As already mentioned, action is understood as the pursuit of goals, the implementation of plans and wishes, and the execution of orders ("par ordre du mufti") until they are realized. Action presupposes perception, recognition and processing of information (cognition) as well as voluntariness, i.e. motivation. Each action proceeds in several steps: Preliminary phase, it concludes an arbitrarily long period of time with the decision on the execution of the action, followed by the control of the course of action. The delimitation of actions takes place through goal setting, which represents the anticipation of the result (anticipation) linked with the intention of realization (intention). In addition to intention (goals), every action also includes cognitive processes. The execution of an action according to instructions excludes intentions and cognitive performances. Of course, the acting person can think about an execution according to instructions. But whether this person then gives preference to his or her own ideas or submits to the instruction depends, in addition to many individual characteristics, especially on the leadership culture, a collective matter, of our society.

In other words, individual action is not a wholly personal matter. The action programs of single individuals are transmitted, aligned and communicated from person to person by the effect of mirror neurons (Franken 2007). So what are mirror neurons again? Mirror neurons are a resonance system in the brain that makes other people's feelings and moods resound in the receiver. What is unique about the neurons is that they already send out signals when someone merely observes an action. An individual's social interactions form an important source for his reactions and decisions. Individuals bear responsibility for the consequences of their actions and include them in their decision making. Thus, every action also has an ethical aspect (Franken 2007).

Different directions of behavioral science use different models of action.

The cognitive model is used as a holistic model of individual action. It views a person as a dynamic unit of action that is shaped by his or her knowledge, is in an active process of exchange with the environment (perception and shaping of the environment through decision-making and [active] action), and is permanently learning (through the consequences of action and internal thought processes) (Franken 2007).

Human action is characterized by voluntary actions, which also allows a certain freedom of decision and action. Seen from a cognitive perspective, human action is a mental process consisting of several phases, which is directed towards changing or maintaining the environment and thus has an actively formative character (Franken 2007).

According to H. Heckhausen (1987), a will formation includes the following phases:

- the reality-oriented motivation phase (readiness to act),
- the formation of intentions (formation and comparison of alternatives),
- the realization-oriented, pre-actional phase (preparation for execution),
- the actional phase (actual action) and
- the postactional phase, which evaluates what has been achieved and takes it into account for subsequent actions (evaluative motivation).

Action can be represented as a chain of the following links: It begins with the motivation to act, followed by the formation of an intention to act (intention), with the crossing of the Rubicon the three actual steps of action take place and ends with the evaluation of the realization. The metaphor "crossing the Rubicon" goes back to the small river Rubicon. Julius Caesar alone owes its fame to this small river. When Caesar decided to cross the Rubicon with his legions on January 11, 49 B.C., and documented his awareness that this was an extraordinarily momentous decision for posterity with the statement "Alea jacta est" (The die is cast), this meant that he had finally decided on civil war and that there was no turning back for him. From now on he single-mindedly did everything in his power to wage and win the war. Caesar finally left the preceding phases of deliberation behind him.

Heckhausen chose this historical event as a metaphor for his model of action, which describes the process from desiring to choosing to willing and acting. The psychological Rubicon is crossed at the transition from choosing to willing, when intentions have emerged from desires and apprehensions and take the form of concrete intentions that are subsequently pursued willfully and determine action (Grawe 2000). "Crossing the Rubicon" in the intentional structure stands for the fact that from then on there is no turning back in the intentional structure. The Rubicon model according to Heckhausen is shown schematically in Fig. 2.4.

Human behavior (or its involvement in individual events) requires another form of explanation.

In order to achieve goals, intentions, purposes, etc., people strive to structure their actions. This form of life management is certainly something that can rightly be called an expression of human reason.

In Hegel, reason is the world principle: "What is reasonable is real, and what is real is reasonable."

Contexts of action are characterised by practical reason; as already emphasised, they are not a purely personal matter, but also a collective one, as can be seen from the three real individual events. In Chap. 4 it will be shown in which contradiction the respective lived leadership culture, a collective matter, can stand to the executing persons.

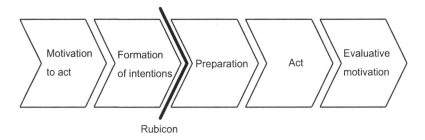

Fig. 2.4 Rubicon of action. Franken (2007) according to H. Heckhausen

The willful action of man is understood as the pursuit of goals, the implementation of plans and intentions in action, he has a certain freedom of choice and action. This freedom is limited by external circumstances but also by the agent himself (his conscience) and his moral principles and instructions ("par ordre du mufti"). In the explanation of actions, the goal and the reasons or other driving force for the action itself must also be specified, as is not done with the cause-effect structure. The sequence of the action also does not directly address the triggering event of the cause-effect structure (see Sect. 2.6.1); instead, the explanatory mechanism for the intentional structure involves crossing the "Rubicon". All three aspects, the freedom of the action, the driving force of the agent and the intentional decision to cross the "Rubicon", are completely different from the elements of a causal chain as it is inherent in the cause-effect structure. All three mental aspects and the intentional crossing of the "Rubicon" are part of a much larger phenomenon, namely, reason. It is essential to recognize that human intentionality can only function when reason is present as a structural, constitutive organizing principle of the whole system (Searle 2006).

At this point it should be emphasized once again how important the distinction is between the causal explanations of the natural sciences and the intentional explanations of the social sciences.

2.6   Application of the Heuristics Found

The causal principle of the cause-effect structure and the volitional actions (intentional structure) are to be described separately for each of the four individual events. Later, the cause-effect structure and its linking principle, the causal principle (connections of cause and effect due to natural laws) and the model of action (intentional structure) are to be combined.

The heuristics developed will be applied to the four individual events presented with the aim of identifying common features that may be useful in answering the question about directing.

2.6.1   Application of the Heuristic for the Cause-Effect Structure

Mario Bunge introduces a metaphor on p. 219 (Bunge 1959), which is taken up because engineers always think in causal terms and it is their task to develop systems with which people's living and working conditions are facilitated. The metaphor of the bow and arrow describes an action that is consistent with scientific determinism.

Exclusively under the aspect of the causal principle, the cause-effect structure of the metaphor describing archery will be considered and not as a description of reasons for action. The dividing line between the two was drawn in Sect. 2.3 (Foundations of intentional structure).

The example used by Bunge is modified and extended.

The order of the four following sections has been deliberately chosen because we believe that it makes the approach to the cause-effect structure clearer.

2.6.1.1 Cause, Generation of an Internal State

The act of releasing a bow (shooting an arrow) is usually considered the cause of the arrow's movement, or better, the acceleration of an arrow, but the arrow will not start moving unless a certain amount (of potential elastic) energy has been stored in the bow by the previous bending of the bow (cause, generation of an internal state) and the arrow has been inserted – as the addition of an external system. The triggering event (relaxation of the bow) starts the process, up to the hitting of the arrow, but does not determine it completely, because the events from the releasing of the arrow to the hitting of the arrow "into the bull's-eye" are faded out with the cause-effect structure.

In order to really be able to identify the tensioning of the bow as a cause and the hitting of the arrow as an effect, we still have to exclude the possibility that there are other factors causally relevant for the effect that correlate with the cause and dominate it in their relevance (Heidelberger 1989). We will talk about this subsequently.

In general, causes are only effective to the extent that they create states through internal processes, in this case potential elastic energy, which enables the externally acting event, the insertion of the arrow, to achieve the intended effect.

2.6.1.2 Entry of an External System and the Triggering Event; Causal Principle

If we follow the example of archery, we must introduce a distinction between cause and triggering event. Through this distinction it is possible to include the basic idea of probabilistic causality in our considerations. Heidelberger M. concluded in the context of his post-doctoral thesis in 1989: "Basically, we are talking about random processes when the effectiveness of the triggering event is completely absorbed by the effect." (Heidelberger 1989).

As already emphasized, the goal orientation of action must not be conflated with causal categories. The imagined goal of action, hitting "into the bull's eye" cannot be a cause; if this were the case, then there would be a violation of the chronological asymmetry of causal reaction, according to which causes always precede effects.

For the firing of the arrow, the cocking of the bow was the cause (generation of potential elastic energy). The drawing of the arrow is the triggering event. The determining causal principle for the process underlying the firing of the arrow is the transfer of potential energy of the bow into kinetic energy (kinetic energy) of the arrow, which the triggering event makes possible in the first place.

The cause, potential elastic energy by tensioning the bow, is an internal state of the system. The decisive change in archery is caused by the addition of an external system – the insertion of the arrow – because without the insertion of the arrow the decocking of the bow would not lead to any detectable result. The point of no return (see Sect. 2.6.1.5) is the decision to release the tension of the bow and thus the prerequisite to trigger the release of the bow, the triggering event is the completion of the decision that enables the transfer of energy. I.e. the following: The initial system under consideration – the tensioned arc – is connected to the triggering event, the execution of the decision, to form a causal chain by the addition of an external system and the decision to release the tension of the arc, the crossing of the point of no return, which is discussed in more detail in Sect. 2.6.1.5; internal processes are therefore not the main source of the process under consideration.

After Bunge:

> The decisive changes are mainly caused by external influences. This means the following: the system under consideration is largely (though never completely) under the influence of the environment; internal processes are therefore not the main source of the changes under consideration. External circumstances acquire their predominance in the way and to the extent that they are able to significantly influence the internal processes. The predominance of external over internal factors is exemplified in technology and industry. (Bunge 1959)

The triggering event unites, so to speak, the inner state of the tensioned bow with the added outer system (inserted arrow). The subsequent flight of the arrow is only made possible with the added external system. Or as Leibniz puts it from a philosophical point of view: "Every passion (inner urge) of a body is spontaneous or arises from an inner force through an external matter (Bunge 1959)".

Successful archery, hitting "into the bull's eye", can only be explained with reference to the probabilistic influencing factors possible for the individual links in the causal chain.

2.6.1.3 Probabilistic Influencing Factors

Let us recall Fig. 1.1, the "Swiss cheese model", of Chap. 1. Thus we see that four parameters can be relevant for the causation of the holes, which could lead to the penetration of the safety barriers and thus to the undesired event.

Note: Fig. 1.1 does not show the necessity of barriers against information decay. But even these barriers are of course subject to the Second Main Theorem of Thermodynamics. This means that the various measures to maintain the functionality of all technical devices can only be counteracted by a corresponding expenditure of energy turnover (maintenance, repair, replacement with newer types of devices, etc.) (Hinsch 2016).

Here, too, probabilistics come into play, i.e., due to the energy consumption continuously required for the operation of technical devices, states can occur for which no barriers were originally intended.

Back to the metaphor of archery: the flight of the arrow is quite clearly dependent on the physical conditions described and is of course not a purely random result. Without the cause-effect structure, hitting "into the bull's eye" would be a pure miracle or, generalized, any technical planning would be a completely irrational behavior. With regard to the probabilistic influencing factors, we must therefore restrict ourselves to those links in the causal chain for which we cannot clearly identify a cause-effect structure. This is shown concretely for the four individual events and below for the metaphor of archery.

Since the beginning of the modern era, the relationship between cause and effect has been understood as deterministic and asymmetrical in time, namely as a relationship that links individual events according to necessary laws.

According to John Stuart Mill (1806–1873), "A cause is a complex of necessary conditions which, taken together, are sufficient for a particular effect to occur." (For the distinction between necessary and sufficient condition, see the corresponding section.) Transferring this to the metaphor of archery: each and every step of the causal chain is fraught with an aspect of probability. A space for probability (a can at the expense of a must) is created. This aspect of probability is passed on, so to speak, to the relationship linking the individual links of a specific cause-effect structure according to necessary laws, by randomly penetrating or failing to penetrate a safety barrier.

Brigitte Falkenburg said:

> Scientific explanations are and remain incomplete. They rely in part on strictly deterministic mechanisms that could also proceed inversely in time, and in part on causal mechanisms that proceed inderterministically and whose causal components obey only probabilistic laws. These explanations cannot uniformly capture the traditional, pre-scientific understanding of causality, according to which the relationship of cause and effect is both deterministic and temporally directed. In the patchwork of contemporary mechanistic explanations, the two traditional aspects of causality are often so interwoven that natural processes are described in sections as reversible and deterministic, and in sections as irreversible and indeterministic. And this is apparently the only way that physics and its successor disciplines can go about getting causal mechanisms that are semi-strict and semi-irreversible. (Falkenburg 2012)

Or, to put it another way, the fact that we perceive parts of the event and/or the whole as random is related to the fact that we do not know the influencing factors in detail and/or that the influencing factors are so extensive that we cannot even grasp them, Bubb (2007), private communication.

With regard to the "probabilistic influencing factors" we restrict ourselves to the flight phase, because the other phases of the bow shot are clearly determined with physical conditions within the framework of the cause-effect structure and only the decision to bring about the triggering event is required, so that the initial impulse created for the arrow by generation of an inner state (tensioning of the bow), addition of an outer system (insertion of the arrow), point of no return, the decision to release the bow and the triggering event (execution of the decision to release the bow) attains the effect, the hitting "into the bull's eye". To regard the arrow as an external system may perhaps be difficult for some readers, because they have the view that the arrow is an immanent part of the system. Here we take the view that there are two systems, the arrow and the bow. We are thinking of the poem by Friedrich von Schiller (1803), "With the Arrow, the Bow," which he put into the mouth of Wilhelm Tell's son Walter, who was playing with a small crossbow.

If the archer aims inaccurately or gives the arrow a slight lateral impulse when releasing it, the arrow will not fly in the intended direction. The archer could, for example, miss the target due to incorrect breathing technique when pulling the arrow. The reverse could also occur, that despite incorrect breathing technique the arrow "hits into the bull's eye". Common sense says that the incorrect breathing technique given as an example is ruled out as the cause of the hit, because normally the probability of hitting the target despite incorrect breathing technique is far less than with correct breathing technique. As life would have it, the incorrect breathing technique at the time the bow was drawn in our example was the cause of the arrow's successful flight. Again the hint, we are thus talking about a random process which completely absorbs the influence of the triggering event on the effect (Heidelberger 1989).

If the three individual events under consideration (Chernobyl, Fukushima Daiichi and Deepwater Horizon) were recurring events – i.e. not unique – non-causality of human behaviour would be discernible. As it is, however, we are left with the limitation of "understanding", which is a foreign body in the cause-effect structure. In accordance with these considerations, causality, the connection between cause and effect, cannot be defined by lawful predictability; it is a human and thus a fallible as well as improvable capacity. The fact that we are able to predict is a consequence of the fact that we know and apply laws of every kind, whereby it does not matter whether they are causal or not.

2.6.1.4 Preliminary Phase

The example of archery must be extended to include a so-called preliminary phase. The duration of the preliminary phase depends on the point of view. If one includes the development of bow and arrow beginning with the age of hunters and gatherers, it covers a very, very long period of time. If, on the other hand, one assumes a spontaneous decision to use the bow and arrow for a specific purpose, the preliminary phase is manageable. Regardless of the point of view, the preliminary phase must by no means be disregarded in the analysis of causes, because the separation of cause and triggering event is not sufficient for answering the question of reliability and direction. In addition, the explanation of the properties of the external and internal systems found requires the inclusion of the preliminary phase.

Kant refers to the term "preliminary phase" introduced here as "predetermination". Predetermination means that, in principle, over any length of time, an earlier state determines by means of laws what happens later. As Kant puts it, "The action is now no longer in my power, but in the power of nature, which irresistibly determines me." (Kant 1793). Recall our observation about directing and writing the script and the role of the actor.

In his work psychological analyses of individual components of work activities, Hacker speaks of "designing action programmes" (Hacker and Richter 2006).

2.6.1.5 Point of no Return

Translated, this means the point from which there is no return or turnaround point, the point of a flight route at which the fuel supply is just sufficient for the return flight. This term is also used in space and aviation. There is a point on a runway beyond which take-off cannot be aborted because the remaining runway length is no longer sufficient to slow the aircraft down safely. A takeoff must be made and, if necessary, an emergency landing attempted.

In other words, a decision must be made immediately before reaching the point of no return regarding aborting or continuing. The decision in the case of archery is: release the bow tension or abort, for whatever reason. The crossing of the point of no return, the decision to release the bow tension, and the triggering event, the execution of the decision, do not occur simultaneously, but in a temporal sequence, even if the temporal differences between the two are as a rule extremely small, as this was proven with the experimental results of Benjamin Libet already mentioned, but still to be presented in detail.

This can mean that the trigger is pulled immediately after the decision to release the bow tension, i.e. the point of no return and the triggering event are almost identical. But it can also mean that the shooter wants to influence his breathing rate in such a way that as little or no cross-impulse as possible can occur when pulling the trigger. As is well known, this is the great challenge in the biathlon, since here the breathing rate has a much greater influence than with archers due to the running effort.

The crossing of the point of no return, the decision to shoot the arrow, always precedes the execution of the decision – the triggering event. How great the time interval between the two is, is determined by the archer. Up to the triggering event no effect, the hitting "into the bull's-eye", would have been ascertainable or that before the triggering event in each case a return would still be possible, if one would recognize at this point that one now crosses the border to the point of no return. In the metaphor of bow and arrow, it is no problem for the agent to perceive, recognize and process the crossing of the point of no return. It is somewhat more difficult in everyday life, for example, when driving in a convoy on the road. There, the detection of the exceeding of the point of no return, the reduction of the safety distance, could be made more difficult, for example, by the sudden approach of a third vehicle. And the detection of exceeding the point of no return becomes even more difficult in the case of such complex events as the three individual events. This is the great challenge; if one were to consciously perceive, recognize and process the exceeding of the point of no return (information processing process), one could intentionally bring about the triggering event if this meant that the intended process could be successfully continued or completed or aborted and thus avoided.

The psychological Rubicon is crossed at the transition from choosing to willing, namely when intentions (intentions to act) have developed from the goals and desires, which are willingly pursued in the further course of action and determine the action.

The point of no return is exceeded in the cause-effect structure when a decision in the sequence of actions, i.e. within an activity, leads to proceeding further, i.e. completing the activity in the belief that the external circumstances or influences will tolerate this excess.

This distinction between crossing the Rubicon and the point of no return is immensely important because of, as has been pointed out several times, the distinction between causes and reasons, purposes, desires, and so on.

There is another difference between crossing the psychological Rubicon and the point of no return of the cause-and-effect structure.

If we imagine Heckhausen's arrow of action, Fig. 2.4, curved into a circle, it becomes clear that we are dealing here with a control loop in which the evaluation of the realized actions in turn leads to changes in the motivational landscape to the left of the Rubicon. So what happens to the right of the Rubicon affects what happens to the left of it, and even more self-evidently, it is the other way around. What happens in the world of desires has a very direct impact on action via goal intentions and intentions (Grawe 2000).

In the cause-effect structure, there are no such feedback effects. Here, once the point of no return has been exceeded, only one thing applies; continue or abort the action. The cause-effect structure knows only one direction, that of the arrow of time of thermodynamics.

For the metaphor of archery, this means that up to the triggering event, the process can be interrupted at any time without the effect occurring. The time difference between the point of no return and the triggering event plays a subordinate role. What is important is that the crossing of the point of no return is brought about by the human decision and always precedes the triggering event, the execution of the decision, in terms of time.

After the triggering event, the probabilistic influencing factors dominate until the effect, the hit "into the bull's eye".

At the same time one can see how one could improve the system (ergonomically): Namely, if one were still able to influence the flying arrow, even if the actual launch failed, one could still direct it to the target. As is well known, with a drone one follows this, Bubb (2007), private communication.

For this interpretation of the point of no return, we refer once again to column driving in road traffic. If, for example, one car is driving too close to the car in front for a given speed and the car in front suddenly brakes for some reason, there is no room for manoeuvre and a collision occurs. In this case, the braking of the car in front is the triggering event, but the point of no return has been created by the car behind because it has shortened the safe distance. Disciplined behavior in this case is a requirement of the trailing driver. The driver in front is expected not to perform any unmotivated braking manoeuvres. If, on the other hand, the driver in front accelerates so that the safety distance increases, the point of no return is no longer relevant. The occurrence of the point of no return depends on both road users, whether the triggering event occurs depends on the preceding driver, provided that the following driver behaves in a disciplined manner on the road after reducing the safety distance. If the person in front continues to drive as before, there will be no collision despite the insufficient distance. This means that the point of no return in column driving is a time-dependent variable that is influenced by time. Distance control systems could ensure that the correct safety distance is maintained.

Our understanding that the point of no return always precedes the triggering event and thus sets the final point, the decisive point, in the chain of events of the cause-effect structure, the further course of which is determined only by probabilistic influencing factors and the effect, should also be underlined with an example from civil engineering.

The keystone is the wedge stone at the highest point (apex) of an arch or the terminating stone in the main node of a ribbed vault. In the arch, the keystone is wedge-shaped so that the horizontal force components give the arch the necessary stability through its own weight. The keystone is effectively a "structural point of no return". Only when it is inserted does the structure become self-supporting and the falsework can be removed (see Fig. 2.5).

This example is intended to show the wide spectrum for understanding the point of no return, but also that the point of no return does not necessarily lead to negative

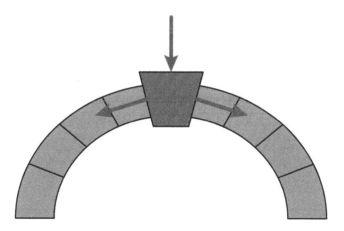

Fig. 2.5 Force diagram for the stability effect of the keystone; "positive" point of no return

consequences, as we found without exception in the events of Chernobyl, Fukushima Daiichi and Deepwater Horizon in Chap. 1. We present an exception to this with the ballad "The Sorcerer's Apprentice".

Summary
The point of no return is the final link in the cause-effect structure connected by the causal chain. Thereafter, the triggering event and the dominant probabilistic influencing factors determine the extent of the effect.

The possibilities for influencing the "point of no return" are shown below on the basis of the three individual events and the ballad "The Sorcerer's Apprentice".

2.6.1.6 Causal Chain in the Case of an Archery Shot

Cause chains, in short, are an approximate model of reality (there are many possibilities, but only one truth).

When explaining a sequence of events in retrospect, one must consider many possibilities, but ultimately that there must have been only one possibility based on what was actually observed or happened – the "truth".

Before we schematize the causal chain of the bow and arrow, let us be clear that we are representing the cause-effect structure here, not the intentional structure. That is, questions such as: Is intention formation finished when the archer has picked up the arrow and the bow, or only finished when he has already strung the bow? What exactly is part of preparation? Does this include taking aim at the target? (Even if that has happened, there is still the possibility of not shooting the arrow) are not the subject of the cause-effect structure, but of the intentional structure. It is not considered here because it deals with purposes and desires, which are not relevant to the cause-effect structure. The cause-effect structure is about causes and not about reasons, purposes, desires, etc., this should be emphasized again.

For archery there is the following causal chain:

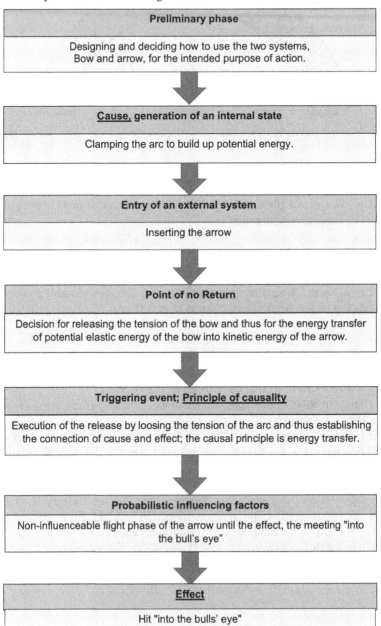

Preliminary phase

Designing and deciding how to use the two systems, Bow and arrow, for the intended purpose of action.

Cause, generation of an internal state

Clamping the arc to build up potential energy.

Entry of an external system

Inserting the arrow

Point of no Return

Decision for releasing the tension of the bow and thus for the energy transfer of potential elastic energy of the bow into kinetic energy of the arrow.

Triggering event; Principle of causality

Execution of the release by loosing the tension of the arc and thus establishing the connection of cause and effect; the causal principle is energy transfer.

Probabilistic influencing factors

Non-influenceable flight phase of the arrow until the effect, the meeting "into the bull's eye"

Effect

Hit "into the bulls' eye"

Cause-Effect Structure

It has been presented in detail for archery using the metaphor above and in the causal chain archery shot.

The crucial thing about this cause-effect structure is that the process could be interrupted at any stage up to and including the point of no return, i.e. before the triggering event, without any effect whatsoever being detected. Only with the decision to relax the bow, to establish the connection principle (energy transfer), was the point of no return passed, the execution of this decision constituted the triggering event, and thus the occurrence of the effect, the hitting "into the bull's eye", whether successful or not, depending on the probabilistic influencing factors, was inevitable.

2.6.1.7 Intentional Structure in the Archery Shot

The archer performs the bow shot according to his intentions. He drew the bow (cause). He placed the arrow (addition of an external system). He triggered the shot with his hand after his decision by relaxing the bow and initiating the flight phase of the arrow. He trusted that all probability-relevant parameters were well-disposed towards him and that the arrow would hit "into the bull's eye".

Each action step of the intentional structure is connected to the corresponding link of the cause-effect structure.

Or to summarize: As a consequence of this example it follows that causal determinism is not complete. Causal determinism says that a cause must be assigned an effect. But a cause can also have different effects (probabilism). In this context, the "butterfly effect" must be addressed. The butterfly effect is a phenomenon of nonlinear dynamics. The butterfly effect is the effect that in the complex nonlinear dynamic deterministic system, a tiny change in the input parameters can lead to unpredictable events in the long run. The following thought experiment is assumed to be the butterfly effect: If a butterfly moves its wings, the resulting vortex of air can nudge a larger one, which nudges another even larger one, and so on. This chain reaction can build up to such an extent that the initially small and harmless flapping of the butterfly's wings arrives on the other side of the world as a tornado.

The question of whether the flap of a butterfly's wings can cause a tornado was posed by meteorologist Edward N. Lorenz in 1972. Common sense, as mentioned earlier, suggests to us that the flap of a butterfly's wings cannot, of course, cause a tornado. However, as is so often the case, data and calculations show that we are quite wrong in our reasoning. Lorenz noted that we can never consider enough parameters to answer the question he posed about a butterfly triggering a tornado ("butterfly effect"). This leaves us with the same problem ("unstirring" the sugar in the coffee cup) as we described in Sect. 2.2 in the digression.

2.7   Causal Chains for the Presented Individual Events

The four individual events are described in detail in Chap. 1. We ask you to remember these explanations.

2.7.1   The Ballad "The Sorcerer's Apprentice"

2.7.1.1 Causal Chain "The Sorcerer's Apprentice"

The ballad "The Sorcerer's Apprentice" is quoted in excerpts in Sect. 1.1. This quotation should be supplemented (The Sorcerer's Apprentice 2018):

> Flow, flow onward
> Stretches many
> Spare not away
> Water rushing …

The ballad tells us the purpose of the action – "Spare not any, water rushin-That to the purpose, water flow" – of the sorcerer's apprentice.

The metaphor Bunge uses excludes the notion of purpose that Aristotle (Falkenburg 2012) superimposed on his types of causes, which we have already addressed.

The ballad "The Sorcerer's Apprentice", although a literary landmark, cannot therefore be used to establish a causal chain.

The ballad is oriented towards the paradigm of human action, it concentrates on the mental reasons for action of the apprentice and not, it should be emphasized again, on the cause-effect structure with which laws of nature are addressed.

With the ballad, Goethe has given literary form to our striving for ever new knowledge, for ever more perfect and comprehensive knowledge. He gives us an explanation of why we cope so astonishingly well with the mental world built up by our brain in our confrontations with the reality around us, which at the same time also addresses the limitations of our cognitive abilities. The human mind has given us deep insights never imagined.

Furthermore, with the ballad "The Sorcerer's Apprentice" Goethe denounced human hubris (the sorcerer's apprentice stands for humanity), which makes people believe that somehow everything can be accomplished. This is countered by our observations. We can observe that nature, if given long enough, overgrows all human activity and slowly, very gradually, strives to restore a "natural" state. Nature must have an inherent memory, as Rupert Sheldracke (2015) argues at length. In physical parlance, memory contents might be called conservation variables. This is very interesting. Memory has something to do with inertia, with persistence (Grawe 2000). Inertia is logically related to cause and effect.

Goethe described to us the connection between science and worldview from his point of view. In order to be able to make the right decisions in each case, not only knowledge is essential, but also firm ethical-moral principles and norms that are part of a worldview (Penzlin 2014).

In addition to the purposefulness, Goethe also addresses the "intentional" error, as it occurs especially with young people (apprentice!), through which the actor wants to confirm himself or seeks to achieve an advantage of whatever kind with minimal effort, cf. Chap. 1 (Der Zauberlehrling 2018). This is also supported by the fact that Goethe wrote the ballad in 1797, i.e. in his Sturm und Drang period, and gave it the title "The Sorcerer's Apprentice".

"The Sorcerer's Apprentice" is also the only event in our observations in which the "Master" was able to avert the catastrophe. Such an intervention of a "Deus ex Machina" (unexpected helper from an emergency) did not occur in the other three individual events, the catastrophes were inevitable there.

For these reasons (deus ex machina and the pursuit of "what holds the world together at its core") we have included the ballad in our consideration of the individual events that led to catastrophic results.

Cause-Effect Structure

This principle is not applicable here. "The Sorcerer's Apprentice" deals with man's inner drive for knowledge-"what holds the world together at its innermost core," see Chap. 1 (The Sorcerer's Apprentice 2018)-and the hubris of humanity. Goethe averts the catastrophe evoked by the apprentice with the appearance of the master.

2.7.1.2 Intentional Structure "The Sorcerer's Apprentice"

"The Sorcerer's Apprentice" shows that in human behavior the respective circumstances (absence of the master) and characters (apprentice) of the persons involved are carriers or triggers for the course of mental processes.

But "The Sorcerer's Apprentice" mainly portrays the urge to transcend limits set for man and to always want to know and explore new things as an inherent trait of man. If man did not have the urge for knowledge and to use this knowledge for his – supposed – good, we would not be able to lead our present life, we would hardly differ from the animal world.

At the beginning of Chap. 1 Prometheus was already mentioned in the quote. Prometheus "the forward-thinking one", bringer of fire and teacher of men. The German philosopher Hans Jonas (1984) opens the preface of his main work "The Principle of Responsibility" with the metaphor of Prometheus unleashed. It expresses how each new technological achievement also brings further evils into the world, like Pandora's box. Pandora's box, as Greek mythology relates, contained all the evils previously unknown to mankind, such as work, disease, and death. The recollection of Pandora's Box represents nothing more than a mythologically transmitted variety of the concept of entropy; the recollection of Cassandra's cries can be interpreted as a symbol of the irreversibility of entropy increase despite warning. Cassandra was never listened to; her unheard warnings are today called Cassandra's cries.

All the evils escaped into the world when Pandora opened the box. The only positive thing the box contained was hope. Before this could escape, the box was closed again. Pandora was part of the punishment for mankind because of the fire stolen by Prometheus. The unleashed Prometheus thus illustrates the threat of a utopian conception of technology in man's modern horizon of power and the urgent need for a new ethic, as Jonas outlines in this work (Jonas 1984). Together with Icarus, the Janus face of this exploratory urge can be clarified. The Icarus myth is generally interpreted to mean that the crash and death of the overconfident man is the gods' punishment for his brazen grab for the sun. The warning of Daedalus not to fly too high or too low, or the heat of the sun or the dampness of the sea would cause him to crash, was disregarded. Icarus became overconfident and soared so high that the sun melted the wax of his wings, whereupon the feathers came loose and he plunged into the sea. This means, on the one hand, that when boundaries are crossed, consequences that cannot always be assessed must be reckoned with and, on the other hand, that there is a spatiotemporal corridor for our actions that must be recognised and observed. A more in-depth treatment of the spatiotemporal aspect of our actions is given in Chap. 3.

These Greek sagas ultimately mean that the ambivalence of human thinking about progress has always been visible. This realization has continued to this day through the many stages of human development.

Prometheus, Icarus and "The Sorcerer's Apprentice" symbolize the threat of an escalating conception of technology with an almost unlimited horizon of human power. "The Sorcerer's Apprentice", like Prometheus and Icarus, knew the consequences of his actions, which is why he also called the "spirits" to the scene, only he did not recognize the point of no return; his countermeasures came too late, after the triggering event.

The intentional structure is clear; it is humanity with its, hopefully ethically restrained, urge to explore and transgress boundaries. The general recognition of limits is not possible in advance. If one is too cautious with regard to possible presumed limits, one achieves no or only very little progress. If, however, one is a little more inquiring in this respect, it can actually lead to progress; one is reminded of the first moon landing, which, in our opinion, involved almost unjustifiable risks. But this ingenuity can also lead to disaster, see the effects of taking thalidomide during pregnancy.

The recognition of limits is ultimately a question of ethics and morality. The German philosopher Hans Jonas (1984) has pointed this out: "The finally unleashed Prometheus, to whom science gives unprecedented powers and economics the restless impetus, calls for ethics which, by voluntary reins, keeps his power from becoming a calamity to man." The sorcerer's apprentice did foresee the effect of his spell. However, he only saw part of it and did not realize the way in which he could return behind the line he had crossed.

To know is to foresee, to foresee is to master; neither was given to the sorcerer's apprentice.

"Knowledge" is the result of human action and work in science and other professions and at the same time a special resource for further knowledge generation, dissemination and benefit creation: knowledge does not wear out through use, but is strengthened, accentuated and further developed precisely through this. The process of knowledge generation therefore does not increase the overall entropy of a system. The application of knowledge can do this very well, especially if, for instance, new methods and techniques of extracting final energy reserves are involved, as in the cases described in Chap. 1. In the area under consideration, increasing knowledge means reducing entropy. However, in principle this can only be achieved by expending energy to achieve this local entropy reduction. This is bought by an increase in entropy elsewhere (where the necessary energy is generated). However, the entire process remains subject to the law of universal decay.

We quote Hochgerner, who in his lecture "The role of social systems in reducing entropy increase" at the symposium "Making sustainability tangible – entropy as a measure of sustainability" in Vienna 2012 stated: "If entropy increase means loss of information (Lewis 1930), one consequence of this is not least social ossification and increase of routines, ritual reinforcement of tradition and growing resistance to change."

More generally: It is a great challenge to see natural events and social systems "thermodynamically". This would mean conceiving of the perpetual competition between structure formation and degradation as an algorithm. The regularities behind the comprehension, which are for the most part still to be explored and probably related to game theory, are effective both in nature and in the social systems of our society.

A digression to game theory: The subject matter of game theory is not limited to games in the common usage of the word. A game in the sense of game theory is a decision situation with several participants who influence each other with their decisions. Decisive for the representation and solution is the level of information of the participants. The connection between entropy and information is discussed at various points in this paper.

The analyses of the following three individual events will show the way in which the natural laws to which actions are subject and the reasons for actions, which are not purely personal but also a collective matter, are intertwined.

2.7.2   Chernobyl (26 April 1986; Explosion of Reactor 4)

2.7.2.1 Chernobyl Causal Chain

For the establishment of a causal chain for the Chernobyl event, we know the most important details; however, there are inconsistencies in this, in particular individual steps in the experimental design and implementation are not clear. For the Chernobyl causal chain below, abstractions therefore had to be made in order to identify a system of order.

Preliminary phase

„The purpose of this was to demonstrate improvements in the capacity of the turbine generators to support essential systems during a major station blackout". (IAEA Safety Series No. 75-INSAG-1)

Author's interpretation: Planning a first test of a new circuit in the event of an unscheduled power disconnection with reactor rapid shutdown, in which the rotational energy of the leaking turboset of the power plant block should continue to be used to drive the main coolant pumps (oversized for post-cooling purposes).

Cause, generation of an internal state

This experiment could only be carried out as part of a planned standstill because of xenon poisoning (see explanation at the foot of this Figure) the reactor no longer has enough slit neutrons ("reactivity") to remove the reactor in the event of a possible test to be able to start again. At the time of the experiment on 26 April 1986, the core rather burnt out, it had only low surplus reactivity and there had to be (by hand) much more control rods to create excess reactivity than allowed. For these the control rods were not designed in operation. They had a precursor with neutron moderator (instead of absorber)

Entry on an external system

Due to repeated requests from the load distributor in Kiev, the attempt to interrupted for about half a day (25 April from 13:05:00 to 23:10:00), neutron-physical behaviour of the reactor was significantly more complex and more confusing (massively under the permissible control rod shutdown reactivity effectiveness, made necessary by xenon poisoning; see explanation at the bottom of this figure). An end to the attempt to at that time would not have had any external impact.

Point of Return

26 April, 01:23:04: Shutdown of the last remaining safety system by to be able to repeat the attempt. "This was a key violation of the test there is no break programme," (IAEA Safety Series No. 75-INSAG-1)

Triggering event: principle of causality

Continuation of the trial: After the interruption started from 23:10:00 continuously increase the reactor capacity to a target value of 700 - 1000 MW (thermal) by pulling the control rods. For quick-switch-off purposes, a minimum number of control rods in the neutron-effective core area (some on "half height"). This provision has been made by the surgeons by active intervention in the control electronics of the control rods (manual ineffective of reactor protection devices). When this state, the rods were connected to the rods by manual quick-off command in the core (approx. 2- 3 seconds) (see the above-mentioned design error of the control rods) and thereby increasing the its reactivity, instead of humiliating, abruptly and led to the non-uncontrolled fission neutrons excursion.

Probabilistic influencing factors

„However, as will be clear from what follows, the accident would not have occurred
but for a wide range of other interrelated events". (IAEA Safety Series No. 75-INSAG-1)
Author' s Interpretation:
1. Consciously exceeding permissible plant parameters by the control room personnel;
2. Organizational failures of management due to insufficient audit of the planned experiment;
3. Complete failure of state oversight and
4. Serious design defects led to the "self-destruction" of an intact enclosure.

Effect

The total destruction of the reactor led to contamination of the environment and to a limitation of the damage, which has not yet been completed.

Explanation of Xenon Poisoning

The distributor of the electrical load in the Ukrainian grid, whose task it is to guarantee a demand-based area-wide provision of electrical energy (load dispatcher; strategic level), had decreed ("par ordre du mufti") that the reduction of reactor power was stopped at 13:05:00 at ~50%. From 23:10:00, reactor power was further reduced to 700–1000 MW (thermal). Due to the ~50% power level reduction for about 12 h, there was still relatively much of the strongly neutron-absorbing fission product xenon (half-life 6.7 h) in the reactor, so that more and more control rods had to be pulled out due to the test interruption. For fast shutdown purposes, a minimum number of control rods in the neutron effective core area (some at "half" height) was not allowed to fall below. The control room personnel circumvented this regulation by active intervention in the control electronics (manual "disabling" of reactor protection devices).

Cause-Effect Structure
The dominant influences for what happens are in the pre-run phase.

> The initiative for the test and the provision of the procedures thus lay with electro technical rather than nuclear experts. The presumption that this was an electro technical test with no effect on reactor safety seems to have minimized the attention given to safety terms. (IAEA Safety Series No. 75-INSAG-1 1986)

Author's translation: The reason for this experiment and the preparation of the regulations for carrying out the experiment lay with the electrotechnical specialists and not with the nuclear specialists. The assumption that this was an electrotechnical test without thermohydraulic and thus neutron-physical feedback effects on the reactor (which are essential in boiling water reactors; the power of a boiling water reactor is regulated via the main coolant flow rate [void effect] and the control rods) led to safety aspects not being taken into account. The Chernobyl reactor type, RBMK reactor (see Chap. 1) is a boiling water reactor by type.

In other words, the experiment was already planned from the outset under inappropriate technical responsibility and under the wrong organisational conditions (strategic level).

Or to put it another way: The essential explanation for the misconduct of the control room personnel is that the experiment was planned by electrical engineers and directed by them. Therefore, repercussions on the reactor were not taken into consideration, the electrical engineers assumed that the reactor would "good-naturedly" and "safely" follow their instructions and the corresponding actions of the operators.

This technically incorrect test planning also illustrates why the actual cause of the largest reactor disaster in the peaceful use of nuclear energy to date was overlooked, namely the design flaw of this reactor type in the design of the control rods. Obviously, the test regulations were not clear either; likewise, no stopping points were provided for checking the test results achieved in the course of the test. In the case of experiments, their failure must be constantly expected and thus the determination of interrogations before a "point of no return" must be obligatory in its run-up.

The reasons for the decision of the line management and the control room personnel (both form the operational level) must be marked with large question marks. The event was triggered by the continuation of the test after an interruption of half a day. Initiated by the lowering of reactor power and the associated unintended formation of too little excess reactivity. Careful situation analysis at the knowledge-based level would have avoided the disaster. The point of no return was set at 01:23:04 on 26 April by the shutdown of the last remaining safety system to allow the test to be repeated. This central violation was set by the test program. This underscores the technical incompetence of the strategic levels. Why the operational level did not veto the determinations within the test programme remains unanswered.

2.7.2.2 Intentional Structure Chernobyl

A successful prediction is known to be confirmed (or disproved) by observation, experiment, or reproduction.

Such a test was to be carried out to control a complete power failure at the reactor. The fact that the planned experimental procedure in Chernobyl was not feasible for this type of reactor was obviously not known. Moreover, it is problematic to regard an experiment as decisive, i.e. as a means of conclusively confirming or refuting hypotheses, since one must never forget that there is nothing definitive in empiricism. Moreover, to claim that successful reproduction is the best test of a prediction is not simply to say that doing entails knowing. If we know how to produce a certain object or phenomenon, we still do not know what it is all about.

Action is not knowledge, it is by no means the source of comprehensive knowledge; it is not the only source of knowledge, even if it provides the best possible confirmation.

Scientific prediction cannot exceed the range of statements of law that form its foundations, and it cannot be more accurate than the specific information on which it is based.

With this (see restriction in Sect. 2.6.1.3 "Probabilistic influencing factors", last paragraph) we "understand" the preliminary phase and the cause, the design error in the control rods (absorber rods) for the reactivity of the reactor.

The addition of an external system – the targeted interruption of the experiment – is extraordinarily complex. Let us recall Fig. 1.1 (the "Swiss cheese model").

Latent and active failure both in line management and through psychological antecedents of unsafe actions among control room personnel has manifested itself in flawed decision-making.

The "deep defence" system of the Chernobyl plant was switched off at the beginning of the second test phase in order to be able to repeat the test. The shutdown of the last remaining safety system is considered a point of no return event. This affects both the control room personnel and many redundant and physically quite different design features.

The continuation of the experiment after an approximately 12-h interruption with a reduction in reactor power represents the initiating event (causal principle). The fact that the triggering event could lead to the penetration of all three protective layers (Fig. 1.1) was due in particular to the atypical system state. Thus, a possible action intervention of the control room personnel was removed from any routine (information decay!).

Taken together, the design error of the control rods (cause), the incorrect interpretation of the safety condition after the approximately 12-h test interruption (addition of an external system), the shutdown of the last remaining safety systems (point of no return) describe, the continuation of the experiment on the instructions of the load dispatcher despite incorrect understanding of the neutron balance (too little control activity) by lowering the reactor power, the triggering event (causal principle) and a number of probabilistic influencing factors which cannot be exactly grasped, the cause-effect structure which led to the self-destruction of the reactor after superprompt-critical power excursion and a contamination also of the further environment.

In the official Soviet reactions to the accident at Chernobyl, there is a tendency to attribute the cause of the accident to personal shortcomings on the part of line management and control room personnel (both of which form the operational level). This must be contradicted on the basis of previous considerations and the fact that design modifications (low-enriched uranium instead of natural uranium and a modified design of the shutdown rods [larger shutdown reserve]) were carried out on the RBMK reactors still in operation, see Chap. 1 (Reason 1994).

In this case, there is an "assumption of competence" by the strategic level, which had left the execution of the experiment to the line management and control room personnel without obtaining a decisional decision from the operational level.

According to all national and international nuclear rules, such a test should have been described in detail in advance, applied for to the supervisory authority and externally assessed. None of this happened. The inadequate state supervision was thus partly responsible for the catastrophic events.

There was no kind of "Deus ex Machina" as we met him in "The Sorcerer's Apprentice", the neutron-physical excursion could not be stopped at any time due to the given constellation.

2.7.3   Fukushima Daiichi (11 March 2011, Destruction of Several Power Plant Units)

2.7.3.1 Causal Chain Fukushima Daiichi

What at first glance appeared to be a technical plant failure triggered by natural events very soon turned out to be a complex event with a very simple cause, in which human, cultural aspects and those influenced by Japanese management culture play a central role.

Preliminary phase
Planning of the power plant without sufficient precaution against external influences (tsunami tidal waves) and internal incidents due to hydrogen explosions.

Cause, generation of an internal state
Construction of the power plant at an altitude of only 10 m above sea level because of (low) energy savings for cooling water pumps; hereto Removal of a natural "hill" by 25 m. As a result, the power plant was now located at approx. 5 m below the height of the highest tsunami waves (see Fig.: 2.6).

Entry on an external system
Two consequences from the Harrisburg incident (Three Miles Island) of 1979 were not hit: passive recombinators (devices that use gaseous hydrogen with oxygen into water) was missing. Similarly, there was no filtered pressure relief of the safety container (containments). although knowledge of their effect on damage limitation. For the plant was thus the zircon-water reaction that occurs in the event of coolant loss unmanageable as a result of radiolysis. The unfavourable arrangement of the fuel element storage basin in the event of earthquakes reinforced the pressure build-up due to increased hydrogen release of the dry fuel elements in the storage tank due to zircon-water reaction (zirconium alloy of the fuel rod enveloping pipe material) and led to a further pressure increase in the containment inertized with nitrogen and (probably after overpressure failure of the containment lid seals) for the release of Hydrogen in the reactor buildings and damage to the reactor buildings.

Point of Return
Nuclear power plants must have an independent emergency power supply. In order to ensure the execution of the security functions even in the event of natural events and influences of third parties (civilization influences). This was obviously not given for tsunami.

Triggering event: <u>principle of causality</u>

On March 11, 2011, at 2:46 p.m. local time, an earthquake of the magnitude
never observed here 9.0 instead.
An effective emergency cooling of the ones in operation at that time
reactor plants 1 to 3 were destroyed as a result of the destruction of the
arranged and unprotected against flooding emergency diesel generators
not possible due to the tidal wave. The tidal wave also put the emergency
power switchgear and the cooling water pumps out of service.

Probabilistic influencing factors

In the last 500 years, a total of 16 (!) Tsunamis with altitudes over 10 m hit the
Japanese coast (incl. Kuril Islands), i.e., around every 30 years. For the northeastern coasts of Honshu, where three continental plates collide, the
geologists believe that only sections (for example 100 km wide) of the Pacific
Plate could simultaneously tear open and slide under the Asian Plate. For this
purpose, the tsunami protection of 5.60 m height was considered sufficient. It
was not assumed that a crack of over 500 km in length with a displacement of
the order of up to 20 m would actually occur. Even more recent experts
reports (2002), which had no longer ruled out this possibility, did not lead to
corresponding consequences, but were attempted to refute by counter-
assessments in agreement between operators, authorities and experts.

<u>Effect</u>

Flooding of the entire plant, destruction of four reactor blocks, radioactive
contamination of a region of about 5 by 40 km in a northwesterly direction.

Cause-Effect Structure

The design of the plant against earthquakes certainly covered the local hazard
potential, adapted over the years to the constantly expanding data base. The reactor
shutdown functioned as designed in all three units (unit 4 was shut down at the time
of the event) at the first earthquake shocks, and although some measured ground
acceleration values exceeded the design in the double-digit percentage range, practically no damage occurred as a result of the earthquake.

On the other hand, the tsunami and thus the emergency power supply design did
not correspond to the state of the art.

The normal power supply feeds all components required for operation and thus
supplies the so-called auxiliary power for all power plant-internal electrically operated consumers (components) of the plant. If no energy is available from outside
through the grid, the generator covers the own demand. The connection to the grid
is disconnected and the power generation of the reactor is reduced to the level of
own demand, so-called island operation, "on-site power". If the on-site power fails,
the reactor protection system shuts down and only the safety systems are still sup-
plied with electrical energy from the emergency power system. The emergency

power system generally consists of redundant diesel generator sets (in Germany eight per reactor unit [four of which are bunkered, i.e. protected against "external effects"]), in Fukushima Daiichi 13 unprotected emergency diesel generator sets for six units.

It is true that the earthquake caused a large-scale collapse of the high-voltage grids in northeastern Honshu (due, among other things, to short-circuits in the substations and knocked-down high-voltage pylons) and thus led to the electrical isolation of the power plant; however, as described above, every power plant should be designed against this, which was obviously not the case.

Thus, after the triggering event, destruction of the emergency diesel generators by the tsunami wave (14 m height), there was no longer any possibility for the control room personnel to intervene (the data refer to Fig. 2.6).

2.7.3.2 Intentional Structure Fukushima Daiichi

The tsunami risk was obviously systematically underestimated. The analysis of the risk to nuclear power plant sites from tsunami waves is a licensing requirement in Japan. For all power plant sites selected at a later date, a higher level was set, typically at 15 m.

Probabilistic models for tsunami risk, which would have included rare severe events, had been developed over the course of four decades, but no consensus was reached on their necessity for application. Instead, the Japanese parliamentary commission of inquiry reported in its official final document that such an organisational failure could only be explained by the cultural peculiarities common in Japan, such as an inability to take criticism, isolationism, persistence despite insight and incompetence (Struma 2006).

We quote from it, "If other people had been in the shoes of the actors, the event would have played out the same way."

Figure 2.6 shows the design and the reactor plant of Fukushima Daiichi and explains the accident described above. It can be clearly seen that the tsunami risk was underestimated.

In particular, the tsunami wave rendered the emergency diesel generators inoperable, so that the emergency cooling of the reactors could not be maintained. It should be emphasised once again that the tsunami wave, as the triggering event, made it evident that the emergency power supply was inadequate in terms of its capacity and state of protection. With the inadequate planning and construction of the emergency power supply, the point of no return was exceeded.

It must be assumed that some of the personnel in the control room were in a state of shock and extreme concern for their relatives as a result of the total power failure – even in the control rooms there was total darkness.

The reactor cores overheated within hours, and the exothermic reaction of the fuel rod cladding material zirconium with steam, which started at approx. 850 °C, led to hydrogen formation.

In this context, the design of the containment must also be addressed. The GE (General Electric)-Mark 1 containment (all four units affected by the earthquake and tsunami wave were so constructed) is not a particularly good solution, either in static or dynamic (earthquake) terms. This is especially true for the fuel pool. The highly arranged water mass of about 3000 t is particularly sensitive in the event of an earthquake. The fuel elements in the storage pool stood dry and provided an additional contribution to the zirconium-water reaction of the reactor core. The hydrogen explosion of the fuel pools destroyed the outer building structure. Figure 2.6 shows the orographic profile of the Fukushima Daiichi reactor plant and explains the accident event described above. It can be clearly seen that the tsunami risk was obviously underestimated.

The relatively high fuel element storage pools also proved to be an unfavourable design in the case of unit 4. Although the unit was not in operation and its reactor was unloaded, a hydrogen explosion also took place there, according to Japanese information with hydrogen from unit 3, which is said to have been transferred via the entire ventilation system. Furthermore, in the days following the explosion, the fuel pool urgently needed structural support because support structures had been damaged.

According to Japanese information, due to the missing cooling of the fuel pools, the fuel levels never dropped below the upper edge of the fuel assemblies, so that no additional hydrogen production is assumed here (not very credible!). In the

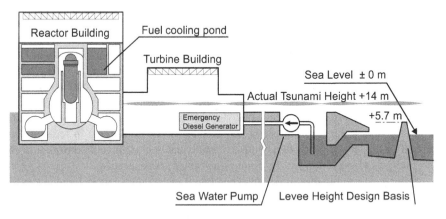

Fig. 2.6 Layout of the Fukushima Daiichi reactor plant with elevation profile. All elevations refer to the reference water level in Onahama Bay, see Chap. 1 (Mohrbach 2012)

meantime the fuel assemblies from unit 4 have been recovered, they are said to be undamaged (except for fallen pieces of debris from the roof; where do you think they came from?).

In the Japanese nuclear industry, cases of falsification and concealment have occurred repeatedly and over many years, and persist even now – see doubts above. These are indications of insufficient acceptance of responsibility for safety (accountability?) and give clues to the typical Japanese shaping of safety culture.

The weakness of people to properly assess perceived risks against other factors becomes evident. The basic psychological mechanisms that help people protect their beliefs and actions, and thus their self-worth, must be addressed. They help him to be and remain capable of acting at all and generally lead to desirable outcomes. In the present situation, these basic characteristics led to undesirable, catastrophic effects, caused by the culture of psychological repression and selective perception that is pronounced in Japan. This strict assessment is based in particular on the following quote from the METI (Ministry of Economy, Trade & Industry) Minister on the occasion of the Ministerial Conference at the IAEA (International Atomic Energy Agency) from 20 to 24 June 2011 in Vienna: "In Japan, we have something called the 'safety myth' (…) it's a fact that there was an unreasonable overconfidence in the technology of Japan's nuclear power generation".

Not to be left unmentioned is the role of "A Hero of Fukushima":

"The then director of the nuclear power plant openly defied the orders of his superiors … The bosses … wanted to stop the pumping in of sea water to cool the damaged reactors …". "The hero refused and continued the cooling on his own authority". (A Hero of Fukushima 2013)

A kind of "Deus ex Machina", as we met him in "The Sorcerer's Apprentice", therefore did not exist.

Summary

The tsunami wave rendered the emergency diesel generators in particular inoperable, so that the emergency cooling of the reactors could not be maintained. It should be emphasised once again: The point of no return was set by the design of the emergency power supply, which was neither sufficient in terms of capacity to supply the safety systems with electrical energy nor did its protection against flooding correspond to the tsunami waves that occurred. The tsunami, the triggering event, hit the unprotected emergency diesel generators with full force.

2.7.4   Explosion of the Deepwater Horizon Oil Rig (20 April 2010)

2.7.4.1 Causal Chain Deepwater Horizon

Preliminary phase
Drilling was 43 days behind schedule on April 20, 2010, the day of the accident, with an estimated additional cost of approximately $30 million by then. Enormous time and cost pressure weighed on the workers of the platform.

Cause: generation of an internal state
In BP drilling, the deposit pressure was approximately 900 bar (the highest deposit pressure encountered to date was 1,750 bar).

Entry on an external system
The blow-out preventer (borehole closure at the head of the borehole) should protect against uncontrolled outbreak of oil and gas was not maintained and the batteries empty. Thus, the hole could not be completely closed. Here it is assumed that the state of the blow-out preventer of the crew on the oil platform was not known. If he had known to her would already be set with the installation of the blow-out preventer the point of no return.

Point of no Return
A hole for a deep hole is drilled in sections: First it goes 500 to 2000 meters in depth. To move the loose rock up from the hole and at the same time cool the drill, a rinsing liquid will come down pressed and rises again loaded with sand and rock. This liquid is much thicker than water, so that the rock can float in it, so to speak, and does not settle. If the predetermined depth is reached, a pipe is inserted that with spacers on the outside keeps a ring space free. There then rises a cement mixture upwards, passing through the inside of the pipe to the sole of the borehole is pressed, exits at the lower end of the pipe and from there the ring space filled up from below. The cement is to stabilize the pipes in the borehole and close the oil deposit absolutely tightly. For this, the cement mixture for deep drilling is much higher quality than in building construction, for example. It contains special additives with which their behaviour can be precisely controlled. Now the crew of the oil rig the Point of no Return: To the cement that made the last piece should fix and seal at a depth of more than 5000 meters, she mixed too much delaying means – it therefore became fixed too slowly. After 24 hours, this was mixture still liquid. The foam cement had the wrong density and a foam stabilizer was not used. But already after 15 hours began the workers with it, the drilling fluid lying above the cement through seawater to replace. The drilling fluid is much heavier than water and would have been so long must remain in the borehole as a counterweight until the cement hardens would have been. Because the deposit presses oil and gas from below with over 900 bar in the borehole- held by the weight of the flushing [Plank 2010].

Triggering event: <u>principle of causality</u>
To replace the drilling fluid above the cement with seawater the drilling crew opened a bubble bottle: Gas shot through the still liquid cement upwards, broke through the insufficient pressure protection at the seabed (blow-out preventer). The gas entered under strong hissing and bubbling out on the sea surface and formed a smog-like spray mist, a high-explosive gas cloud, which covers the 30 meters above sea level oil rig completely enveloped.

Probabilistic influencing factors
Obviously, the spark that led to the explosion of the oil platform came from a ship passing by randomly. It is assumed here that of the oil platform only spark-secured, encapsulated devices are used.

<u>Effect</u>
It was not until August 2010, four months later, that it was possible with the help of a so-called "static kill" to stop the oil inflow and two weeks later via a so-called "bottom kill" to cement the deposit definitively and permanently, to close.

Cause-Effect Structure

The drilling crew tried to complete the drilling as quickly as possible. They therefore decided on an unusual course of action. The crew thus negligently caused the largest oil spill in the history of mankind to date. Deep-sea drilling often encounters oil and gas deposits with extremely high pressures (the highest pressure encountered to date was 1750 bar). Such pressures place the highest demands on equipment (casing wall thickness; blowout preventers, etc.). They require cement slurries with particularly high specific gravity to withstand this reservoir pressure. In the BP well, the reservoir pressure was about 900 bar (Plank et al. 2010). Driven by the enormous pressure at great depth (cause) – about 900 bar prevailed at the bottom of the well – the mixture of oil and gas shot through the semi-solid cement layer and ignited at the surface. On the addition of an external system: In principle, wells are secured against uncontrollable breakout by a closure system. In the present case, a manually operated preventer was used, i.e. in the event of a blowout, an employee had to initiate the closure by hand – unthinkable in the event of an explosion. To make matters worse, it was later discovered that the preventer had been damaged during a previous operation and would not have closed completely even if the batteries had been functioning, see causal chain Deepwater Horizon.

As a cause, the problem of multiple causation, that is, an unconnected plurality of causes and effects, must be addressed. It is less apparent in the case of "united causes", but only becomes apparent when the effect has been produced by each cause on its own, although the joint occurrence of two or more causes does not change anything qualitative about the effect. As interesting as the aspect of multiple causation is, we will not go into it in this paper.

2.7.4.2 Intentional Structure Deepwater Horizon

Hopkins (2012), see Chap. 1, concludes that this was not a "normal accident" (in the sense that Charles Perrow used this term in his book Normal Accidents (Princeton University Press, 1999)), i.e., the cause of the accident can be attributed to a faulty decision-making process rather than faulty technology. Hopkins (2012), see Chap. 1 names in detail for the faulty decision-making process:

- tunnel vision engineering (faulty perception of causality or impaired judgement regarding the wrong cement composition),
- confirmation bias: the well integrity test (the closure of the well was incorrectly considered to be tight),
- falling dominos: the failure of defence in depth (for the technicians, the faulty cement composition was flawless (which was not the case!), consequently the subsequent protective barriers were ineffective (cf. Fig. 1.1, the so-called Swiss cheese model); such a serious error in the philosophy of the principle of the graduated safety barrier concept is unimaginable),
- the meaning of safety (BP was very careful to distinguish between process and personal safety, this went so far as to exclude process safety, especially in drilling),
- process safety indicators and incentives (the BP company used an incentive system focused exclusively on personal safety).

Hopkins' other points are all about management issues.

Overall, it shows that some decision-makers, both at the operator and at the supervisory authority, were not aware of the technical risks of their approach. However, this incident also shows that trust in the supervisory authority is not justified.

The intentional structure is thus simple to present: Man has failed, not only the operator, but also the regulator.

The term "regulator" is understood very differently, even in different sectors of the industry. In some countries, they are state institutions, but they can also be private organisations. There are also constellations in which private and state supervisory authorities work in parallel, for example in the USA. The common goal of all supervisory authorities is to identify and assess the condition of technical objects or, even further, to propose any remedial measures that may be necessary. The sanction options of the supervisory authorities are also very country-specific, ranging from prohibition of further operation to passivity (tokenism).

2.8   Condensation of Individual Events

In Chap. 1 the four individual events of our observations were described. Three of them were catastrophic events in the real world.

With the fourth event, "The Sorcerer's Apprentice", Johann Wolfgang von Goethe created a literary monument for mankind.

Chapter 1 closed with the question of directing.

Two approaches have been used to answer this question, namely.

1. the natural sciences and
2. the social science approach.

The cause-effect structure stands for the natural science approach, the intentional structure for the social science approach.

For both approaches, a strict distinction was made between causes (physical) and intentions to act (intentions). On the basis of this conceptual separation, we have developed the foundations of the cause-effect structure (Sect. 2.2) and the intentional structure (Sect. 2.3).

For the cause-effect structure we have taken as a basis the thermodynamic arrow of time as the connection between cause and effect with its irreversibility.

For the intentional structure, the Rasmussen (1986) cognitive science approach was used.

For both structures, operational approaches were developed using heuristics.

Our heuristic for the cause-effect structure is based on Bunge's (1959) modified and extended metaphor of the bow shot.

For the heuristic of will formation we used the holistic model of individual action according to (Heckhausen 1987 Quoted from Rasmussen 1986).

This approach is, it should be stressed once again, pre-scientific, and that is to say that each of the approaches presented here could prove to be in need of supplementation as the state of knowledge in both the natural sciences and the social sciences increases.

Or in other words, the approaches used are not the philosopher's stone. This idiom is used here in the sense that he is not considered a "panacea". The phrase "the philosopher's stone" (lat. Lapis philosophorum), which is known not to exist, comes from medieval alchemy and goes back to the transformation of base metal into noble metals, especially gold and silver.

After these explanations, we turn to the two structures in detail.

2.8.1 Cause-Effect Structure

The heuristic established for the cause-effect structure states that everything that happens is determined in a lawful way by something else, this something being the external or internal conditions that apply to the object in question (Bunge 1959).

By translating the heuristic to the operational level using the metaphor of shooting with a bow and arrow, the below structure for a causal chain was introduced, cf. Fig. 2.7.

Figure 2.7, the causal chain, implies a detachment (the isolation) of the individual chain links from the overall event and represents an abstraction that is

Fig. 2.7 The causal chain is formed by the thermodynamic arrow of time: Every effect is also a cause

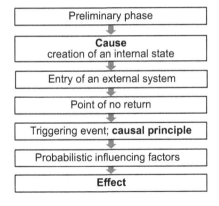

indispensable for the application of the causal concept. Abstraction is not only indispensable for the application of the causal conception, but for all research, whether empirical or theoretical (Bunge 1959).

The cause-effect structure of archery is formed by the conversion of potential energy into kinetic energy.

In the ballad "The Sorcerer's Apprentice" there is no cause-and-effect structure, since no laws of nature are addressed by Goethe.

In the case of Chernobyl, the causal chain is determined by the neutron-physical and thermohydraulic behaviour of the RBMK reactor (boiling water reactor).

The causal chain in the case of Fukushima Daiichi is triggered by a natural event and leads to technical plant failure because no adequate precautions were taken against such external impacts.

As the causal chain in the Deepwater Horizon case shows, this was no "normal accident" that led to the formation of an explosive gas cloud. The chemical energy stored in it led to an explosion pressure wave that destroyed the drilling platform.

It is striking that the causal chain for archery, for the Chernobyl reactor disaster, for the destruction of the Fukushima Daiichi plant by the tsunami wave and the explosion of the Deepwater Horizon drilling platform can ultimately be described by the release of bound energy, i.e. by an increase in entropy (disorder).

Furthermore, it must be noted for the individual events presented that even after the point of no return has been exceeded, the process could have been terminated at any time without any effects being detectable. Only with the triggering event are the links in the causal chain linked to the cause-effect structure. The triggering event itself does not require any energy input in the four cases. Even in the case of the tsunami wave, the energy was stored in the tectonic formation. It can be seen that the overarching cause-effect structure can be understood as a transformation process of energy into different forms, which is accumulated in latent (bound) energy in the system until the point of no return. Only with the triggering event is the accumulated energy released, and the effect is only dependent on the probabilistic influencing factors.

The construction of the cause-effect structure itself is dependent on consequential reasons for action.

We have adopted this term from Julian Nida-Rümelin (2012). We quote from it:

> Consequential reasons for action are directed at causally intervening in the world and bringing about a state of affairs that is different from alternative states of affairs as a result of the action. Consequential reasons for action strive for the corresponding state and use actions (instrumentally) to achieve this change of state.

Background Information
Consequentialism is a collective term for theories from the field of ethics that judge the moral value of an action based on its consequences. Consequentialism is often illustrated by the aphorism "the end justifies the means".

The standard consequentialist conception of action rationality is that a rational person considers the consequences of his actions. If he cannot be certain what the consequences of one action or another will be, he will consider not only the most probable consequence of each action, but also less probable consequences: He will make the evaluation of the action dependent on the probability distribution of its consequences. Not every evaluation of consequences of action is compatible with the reasonableness of the acting person. But a reasonable person will in any case choose an action which, in view of its consequences, seems more desirable than any other possible (open) action in the concrete situation. In consequentialism, the concept of causality is not taken over from the natural sciences and applied to human action, but proceeded the other way round, i.e. causality is assigned to human action. In other words, it is about consequences of action, with which a certain effect is to be brought about by causal interventions. We cannot and do not want to go into the theory of consequentialism in detail here. But it offers an interesting approach to combine laws of natural science with theories of social science. How this could succeed is shown for each specific individual event.

That's what we want to get back to.

Causal interventions, apart from the metaphor for the bow shot, can only be identified for individual process steps in the three catastrophic events. The totality of the cause-effect structure was not the subject of the actions in these events. The situation is different in the case of the bow shot; here the overall process is characterised by consistent intentions, beginning with the drawing of the bow and ending with the effect, the hit "into the bull's eye".

With this understanding, we turn to intentional structure.

2.8.2 Intentional Structure

The consequent reasons for action follow a certain logic, which in the case of the three individual events and the ballad "The Sorcerer's Apprentice" are based on economic or self-centred considerations.

The bow shot uses the causal principle of energy transfer to, perhaps, the intention game for food intake and thus to supply energy to the body, or sporting "hit into the bull's eye", so to meet the intended purpose.

We have already said that Goethe denounces human hubris with the ballad "The Sorcerer's Apprentice". We stay with Mr. Privy Councillor Goethe: "In art and science, as well as in doing and acting, everything depends on the objects being conceived purely and treated according to their nature", cf. Chap. 1 (The Sorcerer's Apprentice 2018). Goethe was not only a brilliant writer, but also a thinker and naturalist. We call his worldview holistic. We have addressed this by showing the dispute between Goethe (world view) and Newton (view of nature) and underlined it by analyzing the ballad. On the other hand, the ballad is also meant to show what happens when a bumbling "sorcerer's apprentice" tries to claim his master's powers by ineffectual means. Goethe has clearly shown us with his famous ballad that we live in a single, tremendously interconnected system. We can describe nothing without the other. Nothing at all. And that is precisely what is called holistic (Lesch 2016).

Or to put it another way: The natural sciences today are pursued quantitatively. And Goethe was not a great friend of quantity. He was a friend of quality (Lesch 2016).

Summary

Goethe's depiction stands for a holistic view of the world. Today we use the procedure of reductionism in the natural sciences. We try to solve problems by reducing them. Here we have gone the opposite way, we have not tried to explain the four individual events, but we have deductively approached the four individual events from the cause-effect structure via the cognitive action approach.

In the case of Chernobyl, the triggering event was the execution of the load dispatcher's order (the load dispatcher is, let us remember, an institution whose task is to harmonize the production and consumption of electricity at the regional level), the fulfillment of which was possible only through active intervention of the control room personnel in the control electronics of the control rods (manual disabling of reactor protection devices). The resulting increase in reactivity led to uncontrollable fission neutron excursion. The directive itself was issued in order to be able to report the fulfilment of the five-year plan. The directive set aside the laws of neutron physics and thermohydraulics for reactor control. Had the directive not been issued, the test could not have been repeated until the next reactor shutdown, perhaps three years from now.

Fukushima Daiichi represents the sudden release of tectonic stress that led to the tsunami wave, the triggering event. The reason for operating the reactors despite their inadequate protection can only be based on economic calculations. Why else have expert opinions and counter-opinions been prepared until the most favourable expert opinion for the management's decision was found? There was certainly a sensitivity on the part of the management to the spontaneous release of tectonic stress, but this was suppressed by the search for a "pleasant" expert opinion.

The aversion to management that shines through here is intentional. With the phrase "par ordre du mufti" we have, so to speak, unrolled the "Ariadne thread". It will continue to haunt us. (Ariadne's thread, after the legendary Cretan king's daughter who enabled Theseus to find his way back out of the labyrinth with a ball of wool: understood here as something that constantly accompanies us).

The crew of the Deepwater Horizon drilling platform was under enormous time and cost pressure, which was the cause of the faulty decision-making processes and which also led to the installation of a non-functional safety device against pressures of approx. 900 bar (reservoir pressure). The operating team failed to operate the blowout preventer correctly. On top of this, the poorly maintained batteries for the closure mechanism failed. To make matters worse, it was later discovered that the blowout preventer had been damaged during a previous operation with the drill pipe and would not have closed completely even if the batteries had been working (Plank et al. 2010).

In the case of the arc shot, the events of Chernobyl, the destruction of the Fukushima Daiichi plant and the explosion of the Deepwater Horizon drilling platform, decisions based on economic considerations formed the intentionality constitutive of action. Economic practice thinks primarily in terms of benefits and costs, weighing expected returns against risks in a limited way.

We Are Heading Straight for the Following Questions
1. What are the connections between the causal chain based on natural laws and the mental processes of action based on the laws of thought?
2. How can the consequentialism of the physical cause-effect structure and the intentional structure of the social sciences be brought together?

2.9   Merging Cause-Effect Structure and Intentional Structure

The conflation requires an explanation that takes up the metaphor of the bow shot with regard to consequent actions, because this does not require any abstractions to be made to the sequence of actions, as is necessary to achieve the necessary transparency in the three individual events.

The archer intends, perhaps, to shoot game. This is a consequent reason for action (causal intervention), because the bow shot is intended to procure food to maintain the metabolism (metabolism) and to eliminate the state of deficiency. However, the archer does not know whether his motivating intention is actually fulfilled with the withdrawal of the arrow. With a certain probability the archer hits the game, then his reasons for action (motivating intention) are fulfilled. With a remaining probability this fulfilment does not occur. The archer has decided, based on his "probabilistic" and "causal" expectations, to release the arrow in order to fulfill his motivating intention. This decision belongs to the type of antecedent intention and is fulfilled by pulling the trigger itself. However, the reasons for action remain unfulfilled until the death of the deer. Only when the death of the deer occurs are mental and causal intentions to act merged. "Probabilistic" expectancy, a somewhat unfortunate term, is used to describe not being certain but assuming that there is a sufficiently high probability that an action will be successfully completed. Contrast this with "causal" expectation. In this case, the agent is quite certain that the intended action will actually lead to the intended goal. As a rule, most people assume a certain outcome of their actions when they initiate an action, especially when the non-success of their action is associated with considerable risks, as was the case with the Chernobyl nuclear disaster and the explosion of the Deepwater Horizon oil rig.

With the death of the deer all conditions of the causal and mental double character of the action of the archer are fulfilled.

The chains of events of the other three individual events also have a double character, which goes back to the cause-effect structure and the sociological logic, here used as intentional structure.

Summary

Intentionality constitutive of action is complex (Nida-Rümelin et al. 2012). Mental intentions can be distinguished into motivating intentions (reasons for action) from preceding (decisions) and accompanying (behavioural control) ones. By pulling the arrow, the archer has fulfilled part of the action-guiding intentionality – that is, causality. With the withdrawal of the arrow the archer undertakes the attempt to kill the game. Thus, part of his decision (antecedent intentionality) is fulfilled. For antecedent decisions, it is important that we have complete control over their fulfillment and thus be able to causally intervene in the event to bring about a change of state. The situation is different for motivating intentions. Here it depends solely on the archer whether the game is killed.

While preceding intentions, i.e. decisions, are fulfilled by the action itself, motivating intentions are only fulfilled by the result of the action. The accompanying behavioral control is a common element for both structures and is also fulfilled by the action itself.

It is about the conflation of action consequences according to the natural law cause-effect structure and the mental intentions known to the behavioral sciences. Overall, we want to give an answer to the question of the direction or reliability of action. This may sound daring. To answer the question of directionality, we must walk a tightrope. In doing so, we have no choice but to combine the cause-effect structure with the causal principle linking them on the basis of the laws of nature and the intentional structure according to the holistic model of action on the basis of free will, which includes all gaps in knowledge (see Sect. 2.10). In short, we assume that a large part of the concrete world we shape "slips through the meshes of the scientific net", cf. Chap. 3 (Whitehead 1979).

With regard to direction, we have adopted the view of Lesch (2016), according to which the four forces, the electromagnetic force, the strong and the weak nuclear force as well as gravitation, stand for the natural scientific explanation of direction. The four forces are beyond any human influence, they are laws of nature that can be attributed to human action.

A clear definition of human reliability is rather difficult. Human reliability needs to be constantly reassessed, as it can change over time. In the human realm, it is not just one particular characteristic that matters, but the interaction of several factors. Reliability or dependability is considered a virtue of character. The legal interpretations of the concept of reliability will not be discussed here. While reliability in technology is a characteristic of technical products.

As a means of influencing human reliability, a wide range of parameters is available in the social science field. We will restrict ourselves here to leadership by management, keyword "par ordre du mufti", knowing full well that a common understanding is necessary for the terms "reliability" and "directive competence". We will deal with this in detail in Chap. 4.

For the metaphor of archery, just as for the individual events of Chernobyl and Fukushima Daiichi as well as Deepwater Horizon, we focused on the cause-effect structure and also addressed the question of intention. For the ballad "The Sorcerer's Apprentice", the existence of a causal chain was denied and therefore only the phases of the action process were shown.

For the consolidation, a joint presentation of the phases of the cause-effect structure and the intentional structure of the action process should precede.

The upper level of Fig. 2.8 shows the cause-effect structure as it was or is shown in Fig. 2.3 (causal principle) and divided into seven parts with the causal chains (arc shot, Chernobyl, Fukushima Daiichi and Deepwater Horizon) event-specific and in Fig. 2.7 (thermodynamic time arrow) event-independent with the individual elements.

The lower level shows the six execution steps of the action as described by (Heckhausen 1987 Quoted from Rasmussen 1986) and shown in Fig. 2.4.

Upper level: Cause and effect structure

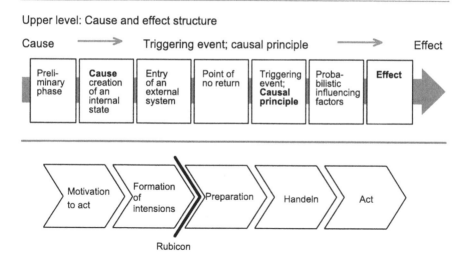

Lower level: Rubicon model of action according to Heckhausen (1997)

Fig. 2.8 Representation of cause-effect structure and the intentional structure (the red line stands for the distinction between natural and behavioural sciences)

The representation in Fig. 2.8 is interesting from a pedagogical point of view, but it should not be understood in such a way that the cause-effect structure is arranged quasi above the level of the action steps. Rather, the two levels are closely inter-locked. The interlocking is formed by action-constitutive intentionality, that is, of causal behaviour and mental intention.

Here, too, we have to limit: This account of the correlation of the cause-effect structure (laws of nature) with the intentional goals of action (laws of thought) is certainly also incomplete, but it plays an important role for explaining events and the related questions about the direction and reliability of human actions, with which we will deal in Chap. 4, as already mentioned.

Back to the distinction between reasons and causes. We have shown why there is a clean distinction between reasons and causes and explained why purposes are reasons and not causes.

In short, we cannot treat laws of thought as laws of nature, for they are laws to be obeyed, not laws to be obeyed, and the physicalist must acknowledge the laws of thought before he can acknowledge the laws of nature. The intentions and purposes of the agent play an essential role on the psychological level in answering the question of intentions. A physical machine, however, can neither demand nor acknowledge anything (Eddington 1931).

Before we interpret the relationships shown in Fig. 2.8 for the three real events and the metaphor "bow shot", starting with the help of the cause-effect structure, one more common feature should be dealt with.

This concerns in the cognitive model of (Heckhausen 1987 Quoted from Rasmussen 1986) the position of the "Rubicon" and in the cause-effect structure the "triggering event; causal principle".

In the cause-effect structure, the course of action remains without effect even after the point of no return has been exceeded. Only with the triggering event are cause and effect linked by the causal principle.

A parallelism to the sequence of actions of (Heckhausen 1987 Quoted from Rasmussen 1986) is present. Here, the crossing of the Rubicon stands for the triggering event that links the motivating with the preceding and accompanying intentions. The initial links in the chain of motivational psychological steps involve transforming intentions into a will to act. Willpower results from the desirability and perceived feasibility of the intentions. Results of action-not consequences, as they arise in the cause-effect structure-become apparent only after the Rubicon is crossed. The Rubicon is crossed when the action can be initiated at the right time and no difficulties challenge the realization of the goal intention. Or put another way: The crossing of the Rubicon corresponds to the state of consciousness of the acting persons and is not directed towards an effect as in the cause-effect structure.

After the crossing of the point of no return, nothing dramatic happens unless the triggering event occurs. When the Rubicon is crossed, it is quite different; once it is crossed, the action simply takes place without any possibility of withdrawal, even if it is objectively nonsensical in the meantime. There is a possibility of intervention after the evaluation of the realized action, as we have already described it with the action band curved into a circle. The problem in the technical system is that one does not immediately recognize the point of no return. If this were the case, there would be an improvement in the safety situation. Only with the crossing of the "Rubicon" (in the mental action model of the law of thinking) as well as with the recognition of the crossing of the point of no return and the subsequent triggering event (in the cause-effect structure) is the actual event started.

Clarification of the necessary and sufficient conditions is also essential for understanding the cause-effect structure, for which the following understanding applies:

Necessary condition: Without them it does not work, "conditio sine qua non" (for example, a basketball game is not imaginable without a basket).

Sufficient condition: necessarily ensures the occurrence of the conditional. Other sufficient conditions can also lead to the occurrence of the event. A sufficient condition in a basketball game would be, for example, the occurrence of two teams, without an opposing team only a "throwing practice" without a game would come about.

Sufficient and necessary conditions for the cause-effect structure as well as consequential reasons for action:

Archery Metaphor
The potential energy of the cocked bow is the necessary condition and the insertion of the arrow the sufficient condition for the cause-effect structure. Neither statement addresses consequential reasons for action on the part of the archer, namely hitting "into the bull's eye" (whatever might be understood by that) or perhaps obtaining food, both of which stand for effect. The killed game leads to a change of state in the archer (hunter) or his relatives by food intake.

Chernobyl
The neutron-physically and thermohydraulically "misunderstood" reactor core, the necessary condition, got out of control during an electrotechnical test, the sufficient condition, because the safety shutdown devices were switched off contrary to regulations. The consequent cause of action, is not addressed by the cause-effect structure, was the test of a circuit for the case of an unscheduled grid disconnection with reactor fast shutdown, in which the rotational energy of the expiring turbo set of the power plant unit was to continue to be used to drive the main coolant pumps. In short, it was a matter of fulfilling the five-year plan and perhaps, unintentionally (?), of increasing the knowledge of plant behaviour for the operating personnel.

The "self-destruction" of the reactor is considered to be an effect.

Fukushima Daiichi
The reactor plant, which was not or insufficiently protected against tsunamis (necessary condition), was destroyed by a seaquake and earthquake followed by a tsunami wave, the sufficient condition. The change in condition of the reactors was brought about by the release of tectonic stress. The consequent cause of action to operate a plant inadequately protected against tsunami is economic; it became an illusion due to the force of nature.

Deepwater Horizon
The enormous pressure (900 bar) of the oil bubble was the necessary condition. The incorrect deep drilling technique and the non-functional closure of the well represent the sufficient condition and describe the cause-effect structure. Consequential

causes of action were in the economic domain. The change in condition on the drilling platform could not be counteracted by the actions of the personnel because there were no or inoperable safety devices.

The cause-effect structure shown is completely different from the explanation of the structure of action. At this point it should be emphasized again how important the distinction is between the naturalistic explanations in the natural sciences and the internationalistic explanations in the social sciences.

The presented unfortunate sequences of events did not arise from a closed cause-effect structure of the agents, because none of the respective agents intended at any time the catastrophic change of state resulting from the sequence of events. The individual action steps are related to specific goals and purposes and are consequential to the intermediate state brought about by the specific action step; integrally, however, the cause-effect structure was hidden by the individual action steps (Müller 1996).

At this point, we would like to revisit the adoption of consequentialist reasons for action coined by Nida-Rümelin (2012) as described earlier. Consequential reasons for action: According to this, the causal concept of cause-effect structure is not taken over from the natural sciences and applied to human action, but proceeded in reverse, i.e., causality is assigned to human action. We have also already argued that consequential reasons for action are aimed at causally intervening in the world and bringing about a state of affairs that is different from alternative states of affairs as a result of the action (Nida-Rümelin et al. 2012). We regard the notion of consequential reasons for action as a linguistic symbiosis of cause-effect structure and intentional structure, which connects the effect caused by the action with intentions characterized by the fulfillment of purposes, desires, goals, and so on. Thus actions, like the behaviour of persons, have the dual character already mentioned.

In the dual character of people's behavior, the behaviorist (behaviorism, social psychological research direction that deals only with objectively observable and measurable behavior) perspective focuses on a person's visible behavior and attributes this behavior to personality and intelligence (Franken 2007).

The dual character of actions consists of the cause-effect structure (see Fig. 2.8) and mental, action-constitutive intentionality (see Fig. 2.4). The complexity of action-constitutive intentionality is illustrated in Fig. 2.4 with the Rubicon of action. This intentionality can be divided into motivating intentions (reasons for action). This motivation is called "motivation to act" in the Rubicon of action (Fig. 2.4) and is located to the left of the Rubicon. Preceding intentions (decisions) initiate "intention formation" (Fig. 2.4). They are also to the left of the Rubicon and include crossing the Rubicon (Fig. 2.4), the central decision to act. Concomitant intentions, considered as behavioral control, are shaped by the three activities in the Rubicon of action (Fig. 2.4); "Preparation", "Action", and "Evaluative Motivation", the actual action after crossing the Rubicon. While preceding intentions, the "intention formation" in the Rubicon of action (Fig. 2.4), are fulfilled by the action and its effect itself, motivating intentions are fulfilled by their consequences and results, of the three action steps to the right of the Rubicon (Fig. 2.4) (Nida-Rümelin et al. 2012).

This requires an explanation, which will be given using the example of the Chernobyl reactor disaster (cf. the Chernobyl causal chain).

The preliminary phase of the cause-effect structure describes the consequent reason for action and, at the same time, the intention constitutive of action: carrying out a test of a new circuit in the event of an unscheduled disconnection of the reactor plant from the electrical supply network with reactor fast shutdown, in which the rotational energy of the expiring turbine set of the power plant unit should continue to be used to drive the main coolant pumps. This involves both entry into the cause-effect structure through consequential reasons for action and motivation to act through action-constitutive intention formation (both are to the left of the Rubicon in the Rubicon model). However, the experimental personnel does not know whether with the generation of the internal state – neutron physical state at the beginning of a planned standstill –, the motivating intention, the fulfilment of the five-year plan, can be achieved by the experiment, this depends in particular on the probabilistic influencing factors. With the initiation of the experiment – addition of an external system (requirement of the load balancer) in the cause-effect structure – a part of the action-constitutive intentionality is fulfilled. The experimental personnel decided to perform the experiment based on their physical knowledge and hoped to fulfill their motivational intention. This decision belongs to the type of preceding intentionality, that is, to the left of the Rubicon. Crossing the Rubicon or the point of no return – shutting down the last remaining safety system – fulfills another part of antecedent intentionality. The accompanying, action-guiding action control is fulfilled at least for the first steps of action; the motivating intentionality remains unfulfilled, as does the consequential reason for action through the explosion of the reactor.

The catastrophic end state can thus be attributed to a deficiency in the accompanying behavioral control (intention).

We want to carry out the conflation of action-constitutive intentionality with the consequential reasons for action also for the remaining three individual events in key words.

Consequential reason for action and motivating intention for the archer (cf. the causal chain bow shot) is the hit "into the bull's eye". Both parts of the preceding intentions, the "motivation to act" and the "intention formation" (cf. Fig. 2.4), are formed by the tensioning of the bow and the insertion of the arrow. The archer has decided to release the bow tension and thus tries to fulfil his motivational intentionality. This decision also belongs to the type of preceding intentionality. With the crossing of the Rubicon or the point of no return – completion of the decision to release the bow tension – the preceding intentionality and the motivating intentionality are fulfilled by hitting "into the bull's eye". The action-constitutive intention and the consequent reason for action are "led" by the accompanying intentions (behavioural control) to the fulfilment of the effect, the hit "into the bull's eye", because all probabilistic influencing factors were "well-meaning".

The logical reason for Fukushima Daichii (cf. the causal chain Fukushima Daichii) was the economic operation of a reactor plant. The motivating intention was fulfilled by the construction of a multi-unit reactor plant. Part of the preceding intentions was fulfilled by the operation of the plant; the other part of the preceding intentions was not fulfilled; no safety-related urgently recommended retrofits were carried out. Here, too, as in the case of the Chernobyl reactor disaster, a lack of accompanying intention becomes causal. The intention constitutive of action and the consequent reason for action were not fulfilled by the destruction of the plant as a result of the tsunami wave.

The logical reason for the Deepwater Horizon drilling platform (cf. the Deepwater Horizon causal chain) was the production of crude oil. The motivating intention was to absorb the enormous deadline and cost pressure. The basis for the decisions leading to the preceding intentions was not fulfilled; the well plug at the head of the well was not functional, and the drilling crew proceeded incorrectly in replacing the drilling fluid above the cement with seawater. Again, a lack of accompanying intention leads to disaster. The intention constitutive of action and the consequent reason for action were not fulfilled by the destruction of the rig due to the explosion of the drilling platform, as a result of an ignition spark.

In all three individual events, there is no convergence between the consequential reason for action (causal intention to act) and the mental goal of action of the specific action steps. Or formulated differently: The dual character of actions addressed by the distinction of natural scientific and social scientific explanation spreads out.

We have now considered the dual character of action from the aspect of causal action and mental intention. We will add the aspect of spatiotemporal behaviour in Chap. 3.

Human behaviour (or its participation in individual events) requires a further form of explanation in addition to the dual character of the action.

Why did the agents continue to follow their action-constitutive intention despite the deviation of the specific action steps from the causal action intention?

In order to achieve goals, intentions, purposes, etc., people strive to structure their actions. This form of life management is certainly something that can rightly be called an expression of human reason.

Paraphrasing Hegel, for whom reason is the world principle, we say; "What is reasonable is real for us, and what appears real to us is then also reasonable". Bubb (2007), private communication.

Contexts of action are characterized by practical reason; they are not a purely personal matter, but also a collective one, as can be seen from the three real individual events.

The willful action of man is understood as the pursuit of goals, the implementation of plans and intentions into action, thereby man has a certain freedom of decision and action. This freedom is limited by external circumstances and other decision-makers, but also by the agent himself (his conscience) and his moral principles. In the explanation of actions, the goal and the reasons or other driving force for the action itself must also be specified, as is not done with the cause-effect structure. With the sequence of action, the triggering event is also not directly addressed; for this, in the explanatory mechanism for the intentional structure, there is the crossing of the "Rubicon". All three aspects, the freedom of the action, the driving force of the agent and the intentional or ordered decision to cross the "Rubicon", are completely different from the elements of a causal chain, as it is inherent in the cause-effect structure. All three mental aspects are part of a much larger phenomenon, namely, reason. It is essential to recognize that human intentionality can only function if reason is present and can be applied as a structural, constitutive organizing principle of the whole system (Searle 2006).

At this point it should be emphasized again how important the distinction is between naturalistic explanations in the natural sciences and intentional explanations in the social sciences.

Now we can consider the second step of the description of Fig. 2.8 on the basis of human intentionality, whose structure is characterized by freedom of action, driving force, and the intentional triggering or directive constraints (Searle 2006).

Intentional Structure:

Archery Metaphor
The freedom of action is necessarily given to the shooter. His driving force consisted in the desire for a hit "into the bull's-eye". The voluntary relaxation of the bow stands for the explanatory mechanism. An instruction to the shooter is to be excluded here.

Chernobyl

It can be assumed that "par ordre du mufti" the line management and the control room personnel have been deprived of any freedom of action. The line management and the control room personnel even begin "in blind obedience" unauthorised interventions in the safety electronics.

One explanation for the driving force can be seen in the adherence to the instructions of the load balancer and thus, overridingly, in the fulfilment of the five-year plan. Both can be seen as the "culture" of the existing organizational system.

Likewise, the reasons for the continuation of the experiment, the deliberately induced triggering event, despite presupposed knowledge of neutron physics (keyword: xenon poisoning), can only be explained by a strong pressure to succeed ("five-year plan"), similar to the events of Deepwater Horizon.

Fukushima Daiichi

The freedom of action to forego the adequate protection against external impacts and the internationally urgently recommended retrofits was taken up. The driving force for this was undoubtedly the Japanese management culture, where accountability is not practiced. By exceeding the point of no return, an inadequate and unprotected emergency power supply, the operating personnel and the entire region were "at the mercy" of their fate. A natural event represents the triggering mechanism. Economic reasons can be cited for the intentional triggering.

Deepwater Horizon

The freedom of action was certainly limited in view of the immense deadline and cost pressure. The driving force for the unusual procedure was certainly the so-called "buying of time". The insufficient knowledge of the personnel of the drilling platform and the supervisory authority can be used as an explanatory mechanism. An unintentional triggering of the gas cloud explosion was caused by an accidental ignition spark, which must always be expected.

Having known the cause-effect structure and the intentional structure for the individual events discussed here, we must ask the questions: What do the two structures tell us; can approaches to answering the question of direction and reliability possibly be gleaned from them?

In the metaphor for the bow shot, the two structures are truly interlocked. Any progression within the cause-effect structure is initiated by the archer, who directs through his consequent action. If, on the other hand, one considers the archer as a component of a warring team, i.e. if one assumes other consequent reasons for action and other intentions constitutive of action, the conditions are no different from those in the other three events presented below.

In the Chernobyl reactor disaster, the cause-effect structure, which can be summarised as "neutron excursion", is completely isolated from the intentional structure. A chain of action following the cause-effect structure cannot be detected at any point in the experiment. Here the full broadside of consequentialism hits: 'the end justifies the means'. On the question of direction, it is necessary to refer to the organizational structure of the conduct of the experiment. Here the answer to the questions is particularly exciting: Why was the experiment continued after an interruption of half a day? Was the pressure to succeed brought about by the "five-year plan" actually decisive for action?

The situation is different at Fukushima Daiichi. Here, figuratively speaking, man has put the millstone around his own neck. He was convinced that the reactor units, which did not correspond to the state of the art in terms of their tsunami design and whose operational management must also be judged sceptically, would be spared from "fate". This failure of man can be regarded as "passive" direction.

The Deepwater Horizon event is virtually the inverse of the metaphor of the bow shot. The archer connects the cause-effect structure with his intention to hit "into the bull's eye". It is the same with the personnel of the drilling platform, only in the opposite direction, there the actors move away from the goal of action, a safe crude oil production, with every step of action. This approach was intended to compensate for the "lost" time and reduce cost pressure. The actors wanted to achieve goals in a certain amount of time, and to do this they had to compensate for the loss of time, i.e. direct the action. Here, too, the motto for consistent action "The end justifies the means" becomes an oppressive reality. The question of what role time plays in human action will be the subject of the next chapter, the focus of which is the spatiotemporal structure.

The four individual events show that freedom of action, driving force and the intentional triggering of events must ultimately be shaped by reason and free will (limited by instructions) and must be seen as interwoven with the laws of nature in order to achieve the desired goals through consistent action. Accordingly, decision-making presupposes an appropriate dosage of certain constraints; if these are overlooked or even pushed aside, as in the present cases, one runs the risk of moving from a state in which accepted and controlled barriers are effective into a state that can no longer be controlled, leading to the sudden destruction of the technical system, as happened.

The picture becomes clearer by directly merging the intentional structure with the cause-effect structure for the three individual events.

The metaphor of archery and the Deepwater Horizon event underline the necessity that cause-effect structure and intentional structure must run in sync if the intended goal of action is to be achieved. This also applies, despite the above limitation (reversed sign), to the Deepwater Horizon event.

The Chernobyl reactor catastrophe falls outside the grid because here the question marks for the intentional structure, restricted by the instructions of the load distributor and by the "five-year plan", predominate.

Fukushima Daiichi represents man's utterly incomprehensible conviction that a well-known natural event, that seaquakes and earthquakes can lead to a tsunami wave, will not occur.

The metaphor of the "bow shot" and the three individual events should be further compressed:

Metaphor "Archery Shot"

Marksman follows the cause-effect structure consistently. The intentional structure is completely synchronized with it. Each link of the cause-effect structure is associated with an action step of the intentional structure. All actions occur at the skill- and rule-based level of behavior. No complications occur, and thus there is no need to cross over to the knowledge-based level. The shooter follows the laws of physics, he consistently implements his physical "world view".

Chernobyl

The strategic level has claimed its enforcement authority over the operational level. The intentional structure of the strategic level overrode the cause-effect structure in the experimental process. The "world view" of the strategic level was obviously shaped by established power structures.

Fukushima Daiichi

Due to probabilistic considerations by the management, the inclusion of the cause-effect structure was lost from focus. Here, failure was virtually "pre-programmed" by the Japanese management culture with its economic "world view".

Deepwater Horizon

Economic pressure led to multiple causation, which was not stopped by professional incompetence on the part of the supervisory authority. The incompetence on both sides also refers to the knowledge of the cause-effect structure. The dominant intentional structure on the part of the operator and the supervisory authority can only be justified by "buying time". Here, too, there is an economic "world view" behind what is happening.

We thus again encounter the influence of action-constitutive intentionality on the cause-effect structure described by Nida-Rümelin (2012).

Back to the distinction between a naturalistic explanation based on the cause-effect structure and the intentional explanations of the social sciences, which underlines the problem of free will. Free will must be fully consistent with the laws of nature. Only then can the set goal of action be achieved. There must therefore be a clear distinction between the problem of free will and the determinacy of natural law.

The Second Main Theorem of Thermodynamics, entropy, stands for determinacy according to natural law, the causal principle, the connection between cause and effect.

We cite Searle (2006):

> We really don't know how exactly free will exists in the brain, if it exists at all. We don't know why or how evolution has given us the unshakable belief in free will. We don't know, in short, how free will could possibly work. But we also know that we cannot escape the conviction of our freedom. We can only act if we presuppose freedom.

We summarize: Ultimately, the question of whether or not we are truly free remains unanswered.

But we cannot completely avoid an answer here, because otherwise we would not be able to explain the three individual events and thus the question posed at the beginning by Dschung Dsi would remain unanswered.

However, we must restrictively say that every explanation presented by us could prove to be inaccurate with a further developed state of natural scientific and social scientific knowledge, i.e. it has a pre-scientific character.

The individual events of the real and the intellectual world mentioned above find their driving force in man's inherent urge to transcend limits set for him and, in doing so, to place an unjustified trust in his own competence, as Goethe clearly shows us in his ballad "The Sorcerer's Apprentice".

We freely interpret Hegel when we note that every limit challenges the attempt to recognize and transcend the limit.

By the deductive approach it could be shown that finding out the cause-effect structure and the intentional structure for each of the four individual events does not answer the question posed by Dschung Dsi to the universe 2000 years ago in general, but it does provide an applicable answer for the question of the direction of the described individual events.

The direction of all three individual events was solely human, even if the destruction of the Fukushima Daichii reactor plant could give the impression that it was a natural event. Before the destruction of several power plant units, the point of no return or the Rubicon had already been crossed with the construction and operation of the units in a tsunami-prone region. This is also confirmed by the following quote:

> The Fukushima nuclear accident … cannot be regarded as a natural disaster … it was profoundly man-made disaster – that could and should have been foreseen and prevented. And its effects could have been mitigated by more effective response. (The National Diet of Japan 2012)

(Author's translation: The Fukushima nuclear accident cannot be considered a natural fate, it was an entirely man-made disaster – against which precautions could have been taken that would have prevented the disaster. And would have mitigated its effects through more effective human responses).

In addition, the operational personnel were restricted in their freedom of action after the flooding by decisions of persons entitled to issue directives and were thus unable to face up to their responsibilities.

To imply irresponsible action would not do justice to the facts of the case. Man was so far limited with regard to the responsibility assigned to him that he did not have sufficient driving power and freedom of action. Also the intentional initiation of his action was taken out of his hands, because he had to inevitably ignore causal laws of nature due to instructions.

The following question hereby suggests itself: Can humans, under these constraints, make extensive use of a heuristic that we use pragmatically as an optimization procedure?

Can we really know whether we have the knowledge necessary to make the right decision and whether we can apply it? We can say that we make our decisions predominantly, at least in everyday life, on the basis of insufficient knowledge. On the one hand, the goal can be the desire for a positive result of action and thus to push back possible opposing arguments, but on the other hand, it can also be characterized by the desire to avoid a feared event. In both cases, however, due to limited knowledge, it is possible that event sequences occur in detail that were not to be expected at the moment of the decision, Bubb (2007), private communication.

Generally speaking, skill-based errors often result from "doing business as usual" without thinking. Errors at the skill-based level, as already noted, result from not accessing the knowledge to respond to change at the right moment. Rule-based errors are quite varied, often characterized by something like "doing things by the book" even when the conditions for following the rule are not or no longer present. For both rule-based errors and knowledge-based errors, however, it is particularly true that there is often no time to think about whether a different rule should be applied or whether a decision should be made on the basis of – and this is a further limitation – limited knowledge. For rule-based errors, there is a lack of knowledge about when deviations from routines occur and in what form they occur. At the knowledge-based level, errors result from changes in the work process for which one is not prepared and which one could not anticipate.

Or to put it another way: In causal constellations, deciding about actions according to mental intentions plays a major role. The decision shows itself in the double character of causal (consequential) and mental properties of the action. The decision establishes a connection between cause and effect, it makes causality effective and thus also generates a coupling of causality and time. This approach is extended in the following chapter to include spatiotemporal logic.

Lack of knowledge and limited freedom of action have caused serious errors of action, which have led to the destruction of the plants. Only in the case of "The Sorcerer's Apprentice" did the master manage to positively influence events.

Summary

Man must always consciously perceive and recognize the limits of his actions as well as process them according to his abilities; in doing so, keep in mind that nature takes its toll for transgressions!

Man is subject to the processes of nature in his actions, and it does not matter whether he recognizes them or not. He must obey them. In the case of conflict, as in the three catastrophic events, nature has always won and man has been the loser.

Thus we are confronted with the question: Can man, under these constraints into which he has maneuvered himself in the events described, comprehensively complete a decision-making process?

2.10 Gaps in Knowledge

We asked ourselves in the previous section, in the context of decisions, "Can you really know if you have the knowledge necessary to make the applicable decision and apply it?"

We have also seen that, in detail, sequences of events occur that were not to be expected at the moment of the decision. So there are two components that we have to consider, on the one hand the insufficient knowledge and on the other hand the temporal development of the event. We will take up the influence of the temporal development of the event in Sect. 3.1. We would like to deepen the aspect of insufficient knowledge on decisions with the term knowledge gaps. The gaps in knowledge already mentioned should also be seen together with the understanding of entropy that has been reached.

More generally, gaps in knowledge occur in many different ways in the scientific field. The drive for knowledge is an indispensable prerequisite for closing knowledge gaps. Gaps in knowledge can be closed ex post (in retrospect), but the ex ante (in advance) dimension means, as we have seen with the three catastrophic individual events, a very risky, usually unjustifiable process. This is particularly evident in the Chernobyl nuclear disaster. Here an "experiment" (see preliminary phase in the Chernobyl causal chain) was carried out and the risk of the experiment was imposed on society; in other words, a "socialization" of gaps in knowledge took place. We are experiencing a similar process with increasing digitalisation; it is progressing without "stops" being made for sufficient testing and proving and for risk assessment.

In Chap. 1, the much-cited "Swiss cheese model" by Reason (1994) was addressed (cf. in particular Fig. 1.1). It is characterised by the fact that man-made barriers are intended to guarantee the process flows he has devised and thus to avoid the socialisation of technical risks. However, since the individual slices of this model are also subject to the changes and decay caused by the increase in entropy and the same also applies to event combinations, it is just possible that in special cases an event "slips through the cheese holes" which should actually be prevented by the man-made barriers. This model is based on the idea that the human (and any biological) organism is a complex system with high information content that deviates considerably from the randomness of thermodynamic equilibrium (Bubb 2007). The idea underlying the model thus relates to the implications of the Second Main Theorem of Thermodynamics (increase in entropy and hence disorder).

Via further approaches coming from the natural sciences, the concept of entropy found its way into information theory, according to which entropy increase can be understood as information loss or responsibility diffusion. Information loss and responsibility diffusion have led to the "human errors" in the present four scenarios, which we referred to as knowledge gaps in Chap. 1. Protective barriers are intended to prevent gaps in knowledge and understanding – i.e. information loss and responsibility diffusion – from leading to the accident events described. Diffusion of responsibility refers to the phenomenon whereby a task that obviously has to be done is not carried out, despite the fact that a sufficient number of suitable persons or bodies are involved and pay sufficient attention to it. The term corresponds to responsibility. Responsibility is the information value of a decision. This is the connection between information decay and responsibility diffusion (Luhmann 2006). Information decay and responsibility diffusion can only be counteracted by constant expenditure of energy supply and thus keep the system in the intended state to fulfill the planned functions (Bubb 2007).

By the way, the word restaurant is a French noun from "restaurer": to restore, to strengthen.

Summary

We have adopted the thesis of the causal principle, "that all events without exception are subject to valid laws in nature" (Kant 1793). We have thus ruled out other interpretations of the causal principle and subscribed to the physical interpretation that every causation consists in a certain form of energy transfer. The laws of entropy, and thus of conservation of energy, establish a general course and time direction for natural processes. The causal principle and the arrow of time given by entropy (see Fig. 2.1) are a fact of human experience. Causal thinking is apparently based on a basic structure of our brain.

In this section we have dealt with the gaps in knowledge from the point of view of the causal principle due to the increase in entropy.

In Chap. 3 we will revisit the gaps in our knowledge and look at how decisions are made from a spatiotemporal perspective.

2.11 Conclusion

Naturally, the explanation of any event is preceded by the description of the event. First, the what, where, and how questions must be answered. The answers to these questions are an essential tool for a scientific explanation of nature, thought and society. Only after that can one successfully turn to the why-question of the behavioral sciences. The why-question clearly refers to intention and thus addresses the aspect of the consequent reason for action. Its answer always represents a model conception of what is going on in the mind of the individual. Since such model conceptions are very different, their answer is correspondingly multifaceted.

Adding the why questions abandons the distinction between the humanities and the natural sciences and thus provides answers that, to be considered scientific, need not necessarily be causal. Science is both descriptive and explanatory; one can distinguish description from explanation, but not separate them. The scientific answer to a specific why question is always precise, objective, and exhaustive when the details of the event in question are well understood (Falkenburg 2012). The answer to the why-question amounts to the identification of causes, an important but by no means exclusive path of scientific explanation.

The why question is also a question of a worldly nature, for example, "Why did this fate befall me?" Coincidence, punishment of God, influence of evil elements, etc. The why question can also open the door to speculation.

A complete cause-and-effect structure determined by the totality of natural laws is a necessary but not a sufficient condition for achieving total predictability, and it would be unfair to blame the world for man's inadequacy to show up with the intentional structure.

In human affairs chance is always present and plays a role not only here but in general. Here is an example: Due to air currents and caused by the "law of gravity", dead leaves fall from the tree. But which leaf really falls cannot be predicted with certainty, and this process consequently appears "random".

Causality, in the strictest sense, does not connect cause and effect, but connects the states of a system at different points in time. Later states are not derivable from preceding causes with necessity. Quantum theory is also indeterministic in this sense. Behind the limited predictability of subatomic states, the same "autonomy" is assumed by which "life" maintains itself in relation to the physical laws of its metabolic processes and distinguishes the processes of consciousness from the neurophysiological processes of the brain (Müller 1996). We take up Müller's "General Systems Theory" and quote from it:

> The indeterminacy of physics is regarded as a sign of an inner relationship between natural, life and consciousness processes; the spheres of reality are not reducible to a physical foundation, they behave "complementarily" to each other. The "new qualities" of life in contrast to its material substrate become the model for the freedom of the will and the spontaneity of action in relation to determined structures. (Müller 1996)

Thus it can be said that the question of direction has been brought closer to an answer by establishing and reflecting the cause-effect structure with the intentional structure, but it cannot be answered conclusively. The causal principle, a special case of the cause-effect structure, is also a factor relied upon by everyone in the context of accountability. The concept of responsibility is completely absent from the realm of natural laws; its clarification is reserved exclusively for the intentional structure. This is by no means a trivial statement. Laws of nature describe the processes in nature that are amenable to observation. Even if one includes in such processes the observed actions of human beings, accountability plays no role in the descriptions. Only when one tries to explain human action (see answering the "why, wherefore question"), does one impute certain purposes, reasons, to the action, both of which are not causes that can be inferred with the why question, but pure intentional purposes of action. Depending on the ethics/morality of the society in which this question is answered, one can assign responsibility to this action. Added to this is the difficulty that it cannot be decided whether a person has acted for the reasons assumed in each case. This cannot be decided with certainty either by him/herself or by any other person (Struma 2006).

This leaves only reason and ethics for the order of actions and for the regulation of power for man's actions.

There is only one world, the world we live in, and we need to explain in terms of natural science and the humanities how we exist as part of it (Searle 2006).

Janich (2006) pointedly highlights the relationship between the natural sciences and the humanities in two mirror-image questions.

Starting from their own (supposedly unproblematic) basic assumptions, the natural sciences ask for the respective other objects in need of explanation: How can there be reasonable reasons in a world of causes, while the humanities ask: How can there be causes in a world of reasonable reasons?

In Chap. 3 an attempt of explanation by means of the spatiotemporal structure is presented.

References

Bartelborth T (2007) Kausalitätskonzeptionen. https://www.uni-leipzig.de. Accessed 10 Mar 2019

Bubb H (2007) Menschliche Zuverlässigkeit und Sicherheit. Ein viertel Jahrhundert im Arbeitskreis des VDI. bubb@tum.de. Accessed 10 Mar 2019

Bunge M (1959) Causality, the place of the causal principle in modern science. Harvard University Press, Massachusetts

Determinismus und Kausalität. www.hfrudolph.bplaced.net/Kausal.html. Accessed: 8. März 2019

Eddington AS (1931) Das Weltbild der Physik und ein Versuch seiner philosophischen Deutung. (The nature of the physical world). Friedrich Vieweg, Braunschweig

Entropie. https://de.wikipedia.org/wiki/Entropie. Accessed: 27. Febr. 2019

Falkenburg B (2012) Mythos Determinismus. Wieviel erklärt uns die Hirnforschung? Springer Spektrum, Heidelberg

Faust Zitate-Eine Tragödie von Johann Wolfgang Goethe. http://www.tcwords.com/faust-eine-tragodie-von-johann-wolfgang-goethe/. Accessed: 14. März 2019

Franken S (2007) Verhaltensorientierte Führung. Gabler, Wiesbaden

Grawe K (2000) Psychologische Therapie. Hogrefe, Göttingen

Hacker W, Richter P (2006) Psychische Regulation von Arbeitstätigkeiten in Enzyklopädie der Psychologie, Themenbereich D Serie III Bd 2 Ingenieurpsychologie. Hogrefe, Göttingen

Hawking SW (1994) Eine kurze Geschichte der Zeit. Die Suche nach der Urkraft des Universums, Rowohlt, Reinbek

Heckhausen H (1987) Perspektiven einer Psychologie des Wollens. Springer, Berlin, pp 121–142

Heidelberger M (1989) Kausalität. Eine Problemübersicht. Habilitationsverfahren an der Universität, Göttingen

Hero of Fukushima. Ein Held von Fukushima (2013) frankfurtrt Allgemeine Zeitung; 11. Juli, S. 1

Hinsch W (2016) Die Freiheit der Wissenschaft. FAZ, 11. Mai 2016, S N 4

Hopkins A (2012) Disastrous decisions: the human and organisational causes of the Gulf of Mexico blowout. Cch Australia Limited, Sydney

IAEA Safety Series No. 75-INSAG-1 (1986) Summary report on the post-accident review meeting on the Chernobyl accident

Janich P (2006) Der Streit der Welt- und Menschenbilder in der Hirnforschung. In: Struma D (Hrsg) Philosophie und Neurowissenschaften. Suhrkamp, Frankfurt

Jonas H (1984) Das Prinzip Verantwortung. Suhrkamp, Frankfurt

Kant I (1793) Die Religion innerhalb der Grenzen der bloßen Vernunft. AAVI, 50. Philosophische Bibliothek. Bey Friedrich Nicolovius, Königsberg

Lesch H (2016) Die Elemente, Naturphilosophie, Relativitätstheorie & Quantenmechanik. uni auditorium. Komplett-Media, Grünwald

Lewis GN (1930) The symmetry of time in physics. Science 71:569

Luhmann N (2006) Organisation und Entscheidung. VS Verlag, Wiesbaden

Mackie JL (1975) The cement of the universe, philosophical books, Bd 16. Clarendon, Oxford

Mohrbach L (2012) Seebeben und Tsunami in Japan am 11. März 2011. VGB PowerTech, Essen

Müller K (1996) Allgemeine Systemtheorie. Springer, Wiesbaden

Nida-Rümelin J, Rath B, Schulenburg J (2012) Risikoethik. De Gruyter, Berlin

Penzlin H (2014) Das Phänomen Leben, Grundfragen der Theoretischen Biologie. Springer Spektrum, Heidelberg

Phiolosophenstüben. https://philosophenstuebchen.wordpress.com/2012/01/22/verstand-vernunft-4/. Accessed: 12. Dez. 2018

Plank J, Bülichen D, Tiemeyer C (2010) Der Unfall auf der Ölbohrung von BP – Welche Rolle spielte die Zementierung? TUM. Lehrstuhl für Bauchemie, München

Prigogine I, Stengers I (1990) Dialog mit der Natur. Serie Piper, München, S 29

Rasmussen J (1986) Information processing and human machine interaction. North Holland, New York

Reason J (1994) Menschliches Versagen. Spektrum, Heidelberg

Scorerre's ApprenticeDer Zauberlehrling (2018). http://www.unix-ag.uni-kl.de/~conrad/lyrics/zauber.html. Accessed: 2. Dez. 2018

Searle JR (2006) Geist. Suhrkamp, Frankfurt

Sheldrake R (2015) Der Wissenschaftswahn. Warum der Materialismus ausgedient hat, Droemer, München

Sorcerer's Apppernice. Der Zauberlehrling (2018). http://www.unix-ag.uni-kl.de/−conrad/lyrics/zauber.html. Accessed 2.Det.2018

Spiegelneuronen. https://www.planet-wissen.de/natur/forschung/spiegelneuronen/index.html. Accessed: 11. März 2019

Struma D (2006) Ausdruck von Freiheit. Über Neurowissenschaften und die menschliche Lebensform. In: Struma D (Hrsg) Philosophie und Neurowissenschaften. Suhrkamp, Frankfurt

The National Diet of Japan (2012) The official report of the Fukushima nuclear accident independent investigation commission. Executive summary. https://en.wikipedia.org/wiki/National_Diet_of_Japan_Fukushima_Nuclear_Accident_Independent_Investigation_Commission. Accessed: 5. Dez. 2018

Whitehead AN (1979) Prozess und Realität aus dem Englischen von Hans Günter Holm. Suhrkamp, Frankfurt a. M

Wingert L (2006) Grenzen der naturalistischen Selbstobjektivierung. In: Struma D (Hrsg) Philosophie und Neurowissenschaften. Suhrkamp, Frankfurt

Spatiotemporal Structure

3

3.1 Perception of the Spatiotemporal Structure

In Chap. 1, the four individual events – the ballad "The Sorcerer's Apprentice", the destruction of the Chernobyl and Fukushima Daiichi nuclear power plants, and the explosion of the Deepwater Horizon oil rig – were described. In the process, the questions of directing, the script and the location of the event, the stage, have crystallized.

We have addressed these questions in Chap. 2. In the same chapter we have explained our understanding of the cause-effect structure and the intentional structure. For both structures we have developed heuristics, procedures for solving problems, based on the structures. With the two structures and the corresponding heuristics we analyzed deductively the four single events described in Chap. 1. In this way, we were able to assign to the sequence of events of the four individual events the links of the cause-effect structure or the intentional structure that we presented. We excluded the ballad "The Sorcerer's Apprentice" from our further considerations because of the absence of the cause-effect structure. We have examined the assignments for the three individual events, the destruction of the Chernobyl and Fukushima Daichii nuclear power plants and the explosion of the Deepwater Horizon oil rig, from different perspectives and "condensed" them with regard to the question of direction. The plot features determining the individual events provide a clear answer:

Man did not act reliably in the succession of these events, because he

- in Chernobyl, the cause-effect structure (neutron population) was set aside on the basis of the instructions ("par ordre du mufti") of the load dispatcher, who is responsible for the distribution of electrical energy in line with demand.

V. Hoensch, *The Chernobyl, Fukushima Daiichi and Deepwater Horizon Disasters from a Natural Science and Humanities Perspective*, https://doi.org/10.1007/978-3-662-65319-7_3

- in Fukushima Daiichi put probabilistic laws above the cause-and-effect structure and followed an intentional structure characterized by complacency embedded in Japanese leadership culture (anonymity of decision-making, where accountability remains in the dark).
- by its actions on the Deepwater Horizon drilling rig, ignored the cause-and-effect structure (in drilling the well) by pursuing its intentional structure (buying time).

Chapter 2 concluded, "There is only one world, the world in which we live, and we must explain, scientifically and spiritually, how we exist as part of it."

Janich (2006), cf. Chap. 2, further sharpened this question on the relationship between the natural sciences and the humanities by formulating:

> Starting from their own (supposedly unproblematic) basic assumptions, the natural sciences ask for the respective other objects in need of explanation: How can there be reasonable reasons in a world of causes, while the humanities ask: How can there be causes in a world of reasonable reasons?

For the scientific approach the basics of the cause-effect structure with the link of entropy as causality were worked out.

The foundations of intentional structure, based on the cognitive science approach of Rasmussen (1986), see Chap. 2, were used to answer the humanities question.

In Chap. 2, we tried not to let the cause-effect structure and the structure of intentional action stand unconnected next to each other with the concept of the "consequential reason for action" coined by Nida-Rümelin (2012), see Chap. 2. In this chapter, both structures will be connected by the spatiotemporal relational framework.

Before presenting our understanding of the spatiotemporal relational framework, we want to present those components that we consider essential for understanding the relational framework we have chosen.

The elements of the spatiotemporal relational framework are described in the following order:

- Spatiotemporal structure of physics,
- Being and becoming, a dynamic approach,
- Spatiotemporal structure in consciousness and
- Consciousness and intentionality.

Or, to put it another way, what answer will the spatiotemporal structure present us with to the intriguing question of the reliability of human action?

We can say that everything is bound to time, even the meaning of words. Those who want to see nature and culture globally cannot overlook problems of time. It even determines our way of thinking. The following two quotes stand for this:

The first quote from "The Magic Mountain" by Thomas Mann:

... "What is the time?" asked Hans Castrop, bending the tip of his nose so forcibly to one side that it became white and bloodless. "Do you want to tell me?" We perceive space with our organs, don't we, with the sense of sight and touch? Fine. But which is our organ of time? Do you want to tell me? See, it's stuck there. But how can we measure something of which, strictly speaking, we know nothing at all, not a single property! We say: time is running out. Fine, let it run down. But to be able to measure it ... wait! In order to be measurable, it would have to run evenly, and where is it written that it does? For our consciousness it doesn't, we just assume for the sake of order that it does, our measurements are just convention, allow me ...

The second quote from "Der Rosenkavalier"; music by Richard Strauss and text by Hugo von Hofmannsthal:

The Marschallin sings:

Time, in essence, ..., time, doesn't change anything. Time is a strange thing. When you live like that, it's nothing at all. But then, all of a sudden, you feel nothing but time. It's all around us. It's inside us. ... Sometimes I hear it flowing – unstoppable. Sometimes I get up in the middle of the night and stop all the clocks ... But you don't have to be afraid of her either. She too is a creature of the Father who created us all.

Both authors discuss the passage of time in nature and the time of consciousness, and that space and time are central categories of human understanding of nature.

Thomas Mann addresses more the fundamental issues related to time and space and how our sense organs process the phenomena associated with them. Hugo von Hofmannsthal, on the other hand, restricts himself to the effects of time, the subjective perception of time and the resulting reactions of man, as well as the measurement of time.

Both quotations together stand for man's everyday dealings with the spatiotemporal structure. Experienced time is subjective time, is consciousness of the present, the past, the future. At the same time we know that only the moment is real, the past is already past and the future has not yet occurred. With measured time we keep to the clock in the daily routine and to the calendar in the annual routine in order to coordinate ourselves in everyday life, cf. Chap. 2 (Falkenburg 2012).

3.2   Spatiotemporal Structure of Physics

From everyday experience we know that the position of a point in space can be described by three numbers – coordinates. An event is something that happens at a certain point in space and at a certain time. Therefore, it can be determined by four numbers or coordinates. Again, the choice of coordinates is arbitrary.

The theory of relativity basically does not distinguish between space and time coordinates, cf. Chap. 2 (Hawking 1994).

What is Relativity? Unfortunately, it is practically impossible, especially in the context of this paper, to define the concept of relativity with a concise formulation that is exact and gives a vivid picture. Let us try linguistically. Relativity means relatedness, conditionality and proportionality or relative validity. According to the principle of relativity, every physical process can be represented in the same way in reference systems moving uniformly in relation to each other. The theory of relativity is a theory founded by Einstein, according to which time, space and mass depend on the state of motion of an observer and are therefore relative quantities.

Let us take a brief look at relativity theory and quantum theory. Both theories partly contradict each other fundamentally in their order structure (Bräuer 2005). While relativity is still strictly continuous, causal and local (continuous: motion continuously without jumps; causal: unambiguous connection between cause and effect; local: causes propagate continuously through space and time, at most at the speed of light signals), quantum mechanics is jumpy, non-deterministic and non-local. The commonality of relativity and quantum mechanics is the unbroken wholeness and coherence of all physical phenomena (Bräuer 2005).

We use the following mathematically abstract representation here:

Time and space appear in the basic equations of relativity almost completely equivalent next to each other and can therefore be unified into a four-dimensional spatiotemporal. Mathematically, however, one is not dealing with a four-dimensional Euclidean space, R^4, but with a so-called Minkowski space, M^4.

The appropriate mathematical description finds the special relativity in the representation given by Minkowski in the Minkowski space named after him, in which space and time merge to the four-dimensional spatiotemporal. Points (events) in spatiotemporal are recorded as world points with contravariant coordinates $x^0 = ct$, $x^1 = x$, $x^2 = y$, $x^3 = z$. Contravariant coordinates are quantities that can be transformed in the same way as the coordinates x^1, x^2, ... x^n of this space in linear homogeneous coordinate transformations in an n-dimensional coordinate space. In Minkowski's reformulation, special relativity becomes a theory of spatiotemporal structure whose structure is formed by worldlines. A world line is the four-dimensional representation of the motion of bodies or the propagation of light rays.

In this space, x and ct have no analogous structures, but, for example, x^1 and ict, where c is the speed of light and i is the "imaginary unit" of the complex number. (Minkowski diagram: With ct instead of formula sign t (from Latin "tempus" [time]) on the time axis, the world line of a light particle becomes a grad with a gradient of 45°).

Space and time are not completely identical in special relativity, but the possibility of thermodynamic behavior remains.

In summary, special relativity is a mathematical construction based on the assumptions that the speed of light is a constant and the associated laws of nature are invariant (unchangeable), see also Chap. 2 (Lesch 2016).

General relativity is based on the more comprehensive general principle of relativity, which also regards accelerated reference frames as equivalent and leads to the inclusion of gravitational interactions.

The structure of time is closely connected with the causal connection of the world. In special relativity, the causal structure of events deviates from the classical form only insofar as, because of the finiteness of the speed of light, only those events which lie within the cone of light (The world-line [path in spatiotemporal] of a particle can only run within the cone of light [Minkowski space].) or on it, can be connected by a causal curve, a "time-like" world-line. However, the temporal sequence of events remains the same in all inertial systems (coordinate system moving in a straight line at constant speed.). Thus, in the spatiotemporal of special relativity, there is never a reversal of cause and effect. Nor can causal chains be constructed that reach events in the past. In general relativity, on the other hand, there are some special solutions.

The apparent flow of time is therefore regarded by many physicists and philosophers as a subjective phenomenon or even illusion. It is thought to be very closely related to the phenomenon of consciousness, which, like consciousness, defies physical description or even explanation and is one of the great mysteries of science and philosophy. Thus our experience of time would be comparable to the quality of the philosophy of consciousness and would consequently have as little to do with reality primarily as the phenomenal content of consciousness, for example, in the perception of the color blue with the corresponding wavelength of light.

This means that we have to revisit the discourse between Goethe and Newton already mentioned in Chap. 2.

According to Goethe, however, light does not have anything directly to do with matter; light is something beyond the world of our conscious experience. It expresses itself in each realm of reality in the manner corresponding to the realm. In matter perhaps in a vibrating motion, on the retina of the eye as a chemical process, in the cortex of the brain as electrochemical neuron excitation, and in consciousness as brightness and color. The nature of light itself, its essence, cannot be captured by mathematical expressions. It can be approached by becoming aware of all the manifestations of light, as Goethe did in his Theory of Colours.

Newton finds that light is composed of colored lights. All statements about the composition of light are only statements about arbitrary mathematical models. In physics, depending on the model, light is regarded as a ray, a wave, a particle, an electromagnetic field or a quantum of action.

It is striking how Goethe's criticism of classical physics finds confirmation through relativity, quantum mechanics and chaos theory.

As explained in Sect. 2.7 of Chap. 2, Goethe's and Newton's theories of colour arise from two incompatible world views, the subjective one, which man cannot escape (Goethe), and the objective rational physical one (Newton).

Therefore we cannot avoid to introduce our scientific world view for the spatiotemporal structure. The objectively rational physical view is captured by the theory of relativity. In tabular form the following representation results.

Relativity: In classical mechanics, physical processes are influenced by infinitely extended space and time, but have no influence on the structure of space and the passage of time; in special relativity, on the other hand, physical processes take place in a four-dimensional spatiotemporal, which leads to the phenomena of length contraction and time dilation; finally, in general relativity, spatiotemporal determines the motion and position of matter and vice versa (Fig. 3.1).

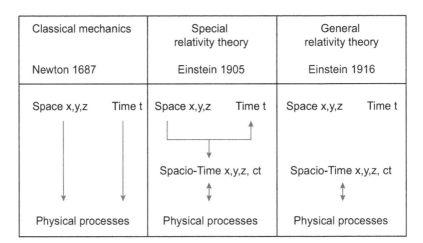

Fig. 3.1 Tabular representation of the relationships between classical mechanics, special relativity and general relativity. (Der Brockhaus 2003)

We have used the same words here that Thomas Mann used in "The Magic Mountain": "... time is running out." Is it correct to say that physical processes "run down" in the four-dimensional space structure? We would be implying that something is happening in time. The spatiotemporal structure is a something that exists. However, we living beings are forced to progress in this structure on the time axis only in one direction, i.e., similar to the fact that there are delimited areas in the space area, in fact there are also delimited areas on the time axis with a beginning (e.g. birth) and an end (e.g. death). These demarcated areas of the time domain are of course valid not only for all living things, but also for all other objects – be it in cosmological dimension, be it in the submicroscopic atomic domain, Bubb (1986), private communication.

However, according to Einstein's Special and General Theory of Relativity, the unit of time can only be realized locally and not globally. It lies in the proper time of moving observers, it does not apply to the entire universe. The measured time is always dependent on the reference system in which the time measurement is made, such as the Earth or the solar system.

Objective, physical time is perspectival. The physical concept of time is an empirically well-supported construct of theoretical physics. The theoretical foundation of physical time is the construction of a uniform time scale ranging from the smallest time unit of Planck time ($5391-10^{-44}$ s) to quite a few billion years from the event of the Big Bang starting to the present world age.

However, the direction of time, i.e. what we would have liked to see explained, still eludes any physical explanation, cf. Chap. 2 (Falkenburg 2012).

Physics cannot explain the direction of the thermodynamic arrow of time, nor the unit of cosmological time.

Let us return to the concept of the arrow of time, which we introduced in Chap. 2 (Sect. 2.2).

Why the broken cup on the floor cannot reassemble and return to the table is explained by reference to the Second Main Theorem of Thermodynamics. According to this, in any closed system, disorder or entropy increases with time. An intact cup on the table represents a state of higher order, while a broken cup on the floor represents a disordered state.

You can easily go from the cup on the table in the past to the broken cup on the floor (present) to the future (trash can), but not vice versa.

The growth of disorder or entropy with time, in the example of the broken cup, is an example of what we call the arrow of time, of something that distinguishes the past from the future by giving direction to time.

Stephen W. Hawking (1994), see Chap. 2, presented his no-boundaries theory at a Jesuit cosmology conference at the Vatican in 1981. He said, "The boundary condition of the universe is that it has no boundaries." He went on to say that the

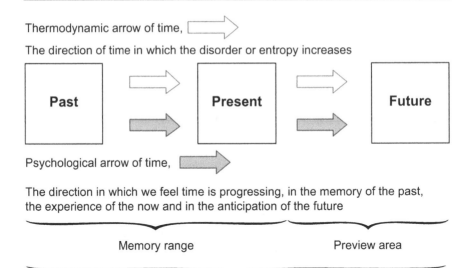

Thermodynamic arrow of time,

The direction of time in which the disorder or entropy increases

Past   **Present**   **Future**

Psychological arrow of time,

The direction in which we feel time is progressing, in the memory of the past, the experience of the now and in the anticipation of the future

Memory range   Preview area

Our experience of time:
Past = memory; **Present** = momentary thought and action; **Future** = Expectation

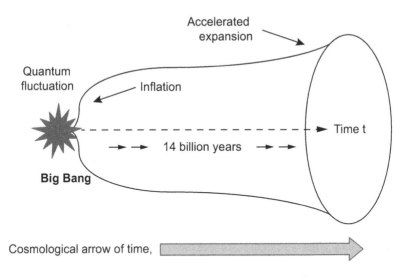

Accelerated
expansion

Quantum
fluctuation   Inflation

Time t

14 billion years

Big Bang

Cosmological arrow of time,

Fig. 3.2 Merging Fig. 2.1 and the Fig. 5.4 "Standard cosmological model of the universe" from Chap. 2 (Falkenburg 2012) with the cosmological arrow of time

proposed no-boundaries condition leads to the prediction that the present expansion of the universe, nearly equal in every direction, is extraordinarily probable. Only the cosmological principle allows us to define a universal, uniform, cosmological time scale.

And with this we come back to the arrows of time already mentioned in Chap. 2, the thermodynamic and the psychological arrow of time, and extend it by the cosmological arrow of time. The term "arrow of time" stands metaphorically (figuratively) for the anisotropy (directionality) of time and cannot yet be adequately explained by the laws of nature.

The cosmological arrow of time results from the expansion of the universe. Edmund Hubbel discovered in 1929 that the universe is not in a static state of rest, but is subject to expansion. The beginning of this expansion is the "big bang" about 14 billion years ago, when the universe was crowded into a single tiny space. Out of this overdense, explosive mixture, the structures of galaxies and star systems have formed. Thus, according to this global measure of temporal asymmetry, the later of two times is the one at which the cosmic structures in question are more expanded. This expansion involves entropy increase (Carrier 2009).

The cosmological principle is thus constitutive for the unity of physical time, for the arrow of time, and for the fact that the present view of the universe makes any sense at all.

We must now, in order to capture the spatiotemporal structure, extend Hawking's considerations presented in Chap. 2 to include the cosmological arrow of time. We ask to take the partial repetition of Hawking's considerations as an underlining.

Only when the thermodynamic, psychological and cosmological arrows of time point in the same direction are the conditions suitable for the development of intelligent living beings. As said before, in order to live, human beings must take in food, which is energy in ordered form, and convert it into heat, energy in disordered form. Therefore, there can be no intelligent life in the contraction phase of the universe. This is why we observe that the thermodynamic and cosmological arrows of time point in the same direction. It is not the expansion of the universe that causes the increase in disorder, but the no-boundary condition causes disorder to increase only in the expansion phase, creating conditions suitable for intelligent life.

The past is connected to the present, as in conventional physics, by causality, but the present is connected to the past by consciousness (Sheldracke 2015). What is Sheldracke telling us with this? Everything present is a moment of experience. When it fades away and becomes a past moment, a new now moment takes its place, a new subject of experience. The just-past moment becomes a past object for a new subject – but also for other subjects.

Or as Whitehead (1979) sees it, mental causality would then act from the future towards the past, while physical causality would act from the past towards the future. According to Whitehead, every current event is doubly conditioned: by physical causes lying in the past and by the self-creating, self-renewing subject who chooses his own past and also makes a choice among possible versions of the future. Through his consciousness, man determines what he brings over from the past into his present, and also chooses among the possibilities that determine his future. Man is connected to his past through a selective memory, and to his potential future through his acts of choice (Whitehead 1979). In view of the most diverse forms of knowledge and representations, various choices for time can be cited, which in their diversity and partial compatibility all refer to the fundamentally subjective-objective construction of human experiences of time.

Even the very smallest processes, such as quantum events, are physical and mental in nature and of temporal orientation. Physical causality moves from the past into the future, while mental events move in the other direction, namely on the one hand by "captures" from the present into the past, but also from potential future variants into the present. There is thus a time polarity between the mental and physical poles of an event: physical causality from the past to the present and mental causality from the present to the past (Sheldracke 2015).

We would now like to return to the question of direction raised in Sect. 2.3 and thus bring the four "elementary forces" into focus. We had established, as did Sheldracke (2015), that all material things consist of quantum particles and that all physical events take place within the framework of spatiotemporal given by the universal gravitational field. We had already addressed the compatibility of the two basic modern theories of physics, quantum mechanics and general relativity. General relativity deals with the macrostructure of the universe – planets, stars and galaxies – and describes gravity, one of the four "elementary forces". Quantum mechanics describes the other three "elementary forces" – electromagnetism and the strong and weak interactions – and is of the greatest accuracy at the atomic and subatomic levels. Special relativity describes relations, the comparison of reference systems moving at uniform speeds relative to each other. General relativity also deals with reference systems that may be accelerated relative to each other. Gravitational fields are accelerated reference systems.

We agree with the view of Lesch (2016), see Chap. 2, according to which the four "elementary forces" are the instruments for directing. We come back to the question of direction in connection with the detection of gravitational waves.

The importance of quantum mechanics is that it explains the elementary building blocks of this world and how to arrange these building blocks into a functional technology.

In short: For the scientific worldview used here, it stands: It is ultimately the thermodynamic arrow of time that writes the script and explains why we have memories of the past, but not of the future. The "time of consciousness" follows the "time of physics (world)" (Carrier 2009).

3.3 Being and Becoming, a Dynamic Approach

By describing being and becoming we want to expand our world view.

Nature progresses creatively and creates qualitatively new things. The traditional notion of a cycle of becoming and perishing was given a directed development by the Second Main Theorem of Thermodynamics. It follows that the future is different from the past in that the world entropically, as it were, "goes deeper and deeper into the red".

What does that mean? Is the difference between past and future, between immutable fact and wavering hope or anxious fear, not much more than an imagination?

Lee Smolin (2015) of the Perimeter Institute also states:

"The future is not real now, and there are no definite facts of it. What is real is the process by which future events are produced."
Smolin goes on to argue that "time – that is, the present and the flow of time – is real. For without this flow there would be no causal relations. In the natural world, everything is real that is real in the moment of time, which is one of many". (Smolin 2015)

Smolin thus believes in a primacy of becoming over being and of processes over structures.

Therefore, this section will address the question of whether there is also a basis in the circulation of nature for the difference between past and future. This difference goes beyond the differentiation between "earlier" and "later". In addition, there is the "now", i.e. a moment of a point in time, which separates past and future from each other and shifts from the past into the future.

Carrier (2009) states, "The question is whether this shift in the "now" corresponds to a natural process. The result is that such a flow of time is essentially tied to the perspective of a consciousness and therefore has no objective counterpart."

All events that are causally connected with each other are in a sequence that is the same for all observers. With this, the transition from the time of physics to the time of consciousness to the "now" seems to concretize itself.

We cite Grünbaum (1973):

The anisotropy or asymmetry of time implies that events form a directed time sequence, so that their objective arrangement according to "earlier" or "later" is possible. Alongside this is the distinction between past and future, which are separated by the present or "now". The now shifts through time and marks the flow of time by its movement. Through the migration of the now, future events continually become present and eventually past.

The boundary between past and future clearly cannot be firmly defined. The now is a fleeting moment in time; no now is privileged, and all pass quickly.

Goethe also dealt with the "now" (Augenblick) in Faust I, verses 1699–1702, Studierzimmer:

"I will say to the moment
Linger! You are so beautiful
Then you may put in shackles
I' wil glady go to ruin!"

So the now has more to do with conscious awareness, and the flow of time says nothing more than that different contents come into the consciousness of an "experiencing I" at different points in time. The now is not a property of physical time, but of an experiencing being. Richard A. Muller is also convinced of this. He states. "What we experience as now is, he says, a moment of new time that is just coming into being. The flow of time is the continuous creation of new Jetzts" (Muller 2016) Rovelli (2016) counters Muller. We quote Rovelli:

But the elementary processes cannot be arranged into a common temporal sequence. On the smallest scale of spatial quanta, the round dance of nature does not obey the baton of a single conductor. Rather, each process dances independently with its neighbor to its own rhythm. The passage of time is part of the world, arises within the world, from the relationships between quantum events that make up the world and themselves generate their own time. (Rovelli 2016)

With the quotation that "the round dance of nature does not obey the baton of a single conductor", we would like to return to the view on the direction taken from Lesch in Sect. 3.2, as well as Fig. 3.2 and the "quantum fluctuation" and "inflation" depicted therein. Could a broader perspective emerge from the evidence of gravitational waves?

100 years after Einstein's work on the theory of relativity, proof of gravitational waves has been achieved. Gravitational waves are caused by accelerated moving bodies. Accelerated masses therefore not only bend space, but they also emit gravitational waves, about whose influence science is still divided. Whether this phenomenon reinforces or modifies our view of direction, that the Very Smallest must fit into the Very Largest, is completely open.

All in all, this means that reality is not as we see it. A still, calm alpine lake is actually a swirling dance of myriads (uncountably large amount) of tiny water molecules. However, Rovelli (2016) also says that not everything about time may be an illusion. He cites at least three aspects to this (Rovelli 2016):

> The temporal continuum: Like other physicists, Rovelli is convinced that time does not pass uniformly, homogeneously and without gaps. It depends on the frame of reference, as relativity has shown, so in itself it is not objective. Becoming: The flow of time is not derived from an exact description of a state, but, as it were, from the point of view of statistics and thermodynamics: "A macroscopic system", such as we, for example, "sees" of the myriads (innumerably large quantity) of microscopic variables (variable quantities) only statistical averages. Only through these statistical phenomena do memory and consciousness come into being, does something like time come into being. Objectively, the "present" does not exist any more than an objective "here". The arrow of time: The direction of time and thus the fundamental difference between past and future must also not be objective.

It is clear to all physicists that the "now" plays a fundamental role in nature.

Zeh (2001) has further clarified this relationship through an apt analogy to the relationship between colors and light:

> The concept of now seems to have as little to do with the concept of time itself as color has to do with light. ... Both now and color are merely aspects of how we perceive time or light ... However, neither their [colors, author's insertion] subjective appearance (such as "blue") nor the subjective appearance of the now can be derived from physical or physiological approaches. (Zeh 2001)

At this point a digression seems necessary, dealing with a possible connection of the reversible (reversible) time of relativity with the irreversible (irreversible) time of consciousness. The irreversible time of thermodynamics no longer refers explicitly to the concept of space, but also to that of time. It speaks of transformation and no longer of motion (de Rosnay 1997). Relativity overturns this understanding. There is a transformation of space into time, because time and space are equivalent. Therefore, one speaks of a spatiotemporal continuum. The time of relativity, however, remains reversible, as in Newton's classical physics. The French physicist O. Costa de Beauregard, in his 1963 book entitled Le Second Principe de la Science du Temps (Costa de Beauregard 1963), presented basic elements by which it is possible to reconcile the reversible time of relativity with the irreversible time of thermodynamics by integrating the facts of thermodynamics, information theory and the physics of relativity.

Back to our worldview about being and becoming.

Beaugard provides a very remarkable hypothesis about the way in which consciousness and the universe are "interlocked" in the dialectical process of observation and action. He uses work by Szilard and Brillouin; according to this, negative entropy (negentropy) and information are equivalent, i.e., he generalizes Carnot's principle. The term negative entropy was coined by Erwin Schrödinger in his book "What is Life?" (Schrödinger 1989) or taken over from Boltzmann (quote from it: "By the way, "negative entropy" is not my invention at all. It is, in fact, the concept around which Boltzmann's independent discussion revolved" (Schrödinger 1989)). He defines life as something that absorbs and stores negative entropy. This means that life is something that exports entropy and keeps its own entropy low: Negative entropy is import and entropy is export. Even if Schrödinger meant free energy by negative entropy, as he wrote in a footnote (Schrödinger 1989), this does not contradict the Second Main Theorem of Thermodynamics, contrary to what is often argued, since this process takes place with the input of energy (in plants, for example, by sunlight).

Or put another way: Information is energy, but a special form of energy that makes it possible to release forces and control processes. Every piece of information is paid for with energy and every increase in energy with information.

We have established that negentropy and information are equivalent. Often information and message are used synonymously. This is not correct. One can receive a message of which one had prior knowledge. In that case it has no novelty value and therefore no information. A message is associated with information only if, prior to the receipt of the message, there was some uncertainty about the subject matter conveyed, which has been removed by the receipt of the message. According to Shannon (founder of information theory), information is not an objective quantity, but depends on the level of reference taken as a basis by the investigator (i.e. subjectively), cf. Penzlin (2014) in Chap. 2. Many physicists identify information with knowledge and entropy with non-knowledge, thus correlating information gain with negentropy. "Just as the information content of a system is a measure of the degree of order, the entropy of a system is a measure of the degree of disorder; and the one is simply the negative of the other," Penzlin (2014) in Chap. 2 quotes Norbert Wiener. Information is order, organization, the non-probable, and thus the antithesis of entropy, which is disorder, dissolution, decay, the probable. Entropy is the measure of the lack of information about a system. Entropy and probability, as we have shown, are closely linked by statistical theory. Comparing the different mathematical expressions, we find that information is the inverse of entropy. It corresponds to an anti-entropy. That is why the above mentioned expression negative entropy (negentropy) was introduced. Thus, information and negentropy are equivalent to potential energy. We repeat: Information is energy, a special form of energy that enables us to create and control processes. This close connection between energy and information was understood when it was realized that energy must be expended to obtain information and that, on the other hand, information must be evaluated to gather and use energy in a controlled manner. Every information is paid with energy and every increase of energy with information. Entropy is the measure of a system's lack of information, information decay, and diffusion of responsibility. Information is therefore the equivalent of negative entropy. Every experience, every action, every acquisition of information by a brain consumes negative entropy. So you have to pay a "fee" to the universe: irreversible entropy. Or to put it another way: gain in information always means loss in entropy, so must be paid for by entropy increase elsewhere. The brain can also create negative entropy and thus increase the organization, order, and amount of information about the system in which it is located; but the whole remains subject to the law of universal decay (de Rosnay 1997).

Note: "The final word on whether both entropy relations, Shannon's and Boltzmann's, actually correspond to each other or are only formally analogous is probably not yet spoken." see Penzlin (2014) in Chap. 2.

All living things, including humans, are on the "road of no return" (de Rosnay 1997) or, in other words, in a process of adaptation. And this adaptation is the hard condition of survival, because we can only affect the environment to the extent that we pick up information from our environment. But man can only observe the phenomena in the sense that he dissolves them, since every acquisition of information is connected with an increase in entropy. Without information everything creative would not be possible. In the four-dimensional spatiotemporal structure, only consciousness progresses because it informs itself. But as a result of the adaptation of consciousness to its environment, consciousness can only explore its environment in the direction of increasing entropy (the direction of "time"). Negentropy is completely neutral and objective. For consciousness, every piece of information has a meaning, a significance, a different subjective value (Fig. 3.3).

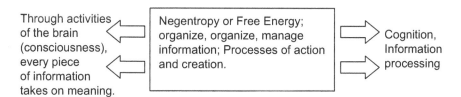

(Free energy is a thermodynamic state variable of a system. In a reversible isothermal process, free energy is the part of the internal energy of the system that can be released to the outside as work in the event of a change of state. The rest of the energy, the bound energy, is not convertible into work.)

Fig. 3.3 The two basic activities of consciousness

As is well known, the concept of information does not exist in physics. It is unfortunately true that the concept of information is used in the literature with very different content and reference, on the one hand in the sense of classical information theory with signal and message, in other cases with negentropy, "meaning" of a message or also with "specificity" or par excellence with knowledge, cf. Chap. 2 (Penzlin 2014). We quote from Penzlin in Chap. 2 (Penzlin 2014): "That a signal becomes a message requires the participation of an individual possessing consciousness … to whom this signal gives cause to decide for a certain of the possible behaviors, to react."

We can summarize by defining information as the content of a message that triggers an action. But information must not be equated with meaning. In this respect, Shannon's concept of information differs from the colloquial one, for which information always means information about something, i.e. it has a content.

Consciousness, we have noted, is thus the linchpin of the dynamic approach to being and becoming. At the same time, consciousness is one of the last great mysteries that science is grappling with. Today we consider consciousness not as a substance but as a property. We have worked out in Chap. 2 that our scientific knowledge does not alone provide us with the preconditions for shaping a "problem-free" social coexistence. We also have the responsibility to use our free will responsibly. For classical determinism, free will seems "impossible" in the scientific realm, whereas the observation of nature is self-evident. That is, there are two basic activities of consciousness. One corresponds to the transformation of negentropy into information. This corresponds to the process of consciousness in which information means "acquisition of knowledge". The other corresponds to the reverse transformation of information into negentropy. This is the process of acting and creating, in which information means "organizing capacity" (shaping). On one side the brain informs, on the other it organizes, shapes. The first process does not cost much negentropy, it deals with measuring and observing nature. The other, reverse process of creative activity costs information.

This figure shows two complementary forms of means of communication; descending and ascending information. Both forms involve the projection of the two basic activities of individual consciousness onto the social level: observation (with the aim of acquiring knowledge, of informing oneself) and the creative act (with the aim of organising the world, of shaping matter) (de Rosnay 1997).

Society has created a mass communication system based on the rapid dissemination of information. This is descending information that travels from the top of the social organization to its base. One of the most important advantages of this information system is that it provides a social feedback loop. Without this feedback, there can be no participation; much less a social community with an internal bond.

But also the temporal difference between these two kinds of consciousness activity (information $\leftrightarrow$ negentropy) is important, because with it the actual time, the "now" is addressed. The actualization time is, so to speak, a moment that we have already dealt with and thus can conclude our dynamic approach of being and becoming.

In short: The arrow of time is part of the natural process, since ordered states are less probable than disordered ones. The arrow of time of the probabilistically interpreted entropy of physics points to states of low order. For given conditions, the maximum of entropy corresponds to the state of least structuredness and highest disorder. Ultimately, as noted, all of these determinations converge in the thermodynamic arrow of time. The path of time runs from order to disorder. And this is based on the fact that the universe as a whole follows this path. It is the irreversibility (irreversibility) thus determined that forms the basis of the anisotropy of time.

The conclusion is therefore that the flow of time remains bound to the human world of experience and thus to consciousness and cannot be explained by physical laws.

For our further considerations there is thus the statement that space and time are central categories of the human understanding of nature.

3.4 Spatiotemporal Structure in Consciousness

After the foundation for our world view is cast, we want to deal with the "organ" that is responsible for the formation of a world view, the brain with consciousness. We have already mentioned both several times without treating them in detail.

A brief digression or background information is needed on the term worldview.

A scientific worldview, repeatedly praised by some ideologues, such as Marxists, is a "contradictio in adjecto" (contradiction in the added). There is no such thing. There are only scientific worldviews.

As a physical system, the brain is subject to probabilistic laws and physical laws. Laws are the description of phenomena by means of mathematical models, that is given in both cases. Ultimately, however, there is only one reality and only one set of laws in this world. If we assign reality to different laws for the sake of better manageability, it is a separation made by man. Perhaps the essence of consciousness can be summarized – not defined – as follows!

People live in social communities as acting individuals. In doing so, they learn to intersubjectively coordinate their subjective experience of time and their actions, which includes time measurement as well as language and the rules of action, it can also be omissions in their social community, cf. Chap. 2 (Penzlin 2014).

We know, from the previous sections, that consciousness plays a causal role in determining our behavior. We have also pointed out that we do not know whether free will and consciousness exist at all. There is considerable doubt in certain areas of science as to whether free will, like consciousness, even exists. This is a dispute that has not yet led to a definitive conclusion. Obviously, scientific observations and (possibly also very old) philosophical considerations are in conflict here.

Since almost all physical processes follow fixed laws, it is possible that the processes in the head are also always determined, i.e. fixed, by preceding neuronal processes. This would mean that the result of an action plan or decision would already be fixed before it occurs. Can we still speak of free will in this case? We also know that free will is a reality provided some crucial assumptions about the role of consciousness, which we will come to.

In certain areas of science there is considerable doubt as to whether free will, like consciousness, exists at all. This is a dispute that has not yet led to a definitive result. Obviously, scientific observations and, perhaps, very old philosophical considerations are in conflict here.

According to the deterministic view, every event has causal causes. If one knows the current state of a system, one can (theoretically) predict all future ones. In reality, this is often impossible because the boundary conditions to be taken into account are too complex and cannot be determined exactly (see, for example, weather forecasts).

The commonly held notion of free will corresponds to the addressed consequentialism, i.e. one can choose between alternatives of action, keyword: "the end justifies the means". This principle is not compatible with determinism. The principle of free will demands that the action is in harmony with the motives and convictions of the person and that it is autonomous, i.e. not under compulsion.

In classical physics, strict determinism applies. Exceptions to determinism are found in quantum physics. Even with complete knowledge of the current state, it is generally not possible to specify the subsequent state exactly. Only probabilities can be given for this.

We quote Herrmann:

Nevertheless, it is true that our will is (potentially) free; free from chance, free from coercion, but conditioned by our experiences. Therefore, determinism does not invalidate the concept of free will, but is, on the contrary, its necessary precondition. The question is not whether our choices in a situation are predetermined, but by what. (Herrmann 2009)

Back to the representation in Fig. 3.1 of space and time, which in four-dimensionality are not separate coordinates, but in Einstein's theories of relativity are united in a continuum or, even further, form an inseparable unity.

For the description of spatiotemporal relations one needs four dimensions, namely three for space and one for time. There are the mentioned four elementary forces: gravity, electromagnetism, strong and weak interaction. Space is a very elementary basis for our conscious experience.

Consciousness is based on perceiving, recognizing and knowing, i.e. on cognitive abilities. Man forms concepts in his consciousness, and so he becomes aware of the world with these concepts. Consciousness develops in the confrontation with the everyday problems and the great ones of the world. Basically, it is always about the struggle for survival, in the biological and/or social realm. The root of consciousness is the confrontation with the problems of this world. By experiencing himself as an individual, man has to face the struggle for survival and solve the ever new problems by making decisions. In the process, consciousness and culture develop (Bräuer 2005). We quote Kurt Bräuer (2005):

> Physics is certainly an important result of the development of human consciousness and a very important foundation of our culture. Physics is essentially based on recognizing and mathematically describing individual parts of the world and their relationships to each other.

Space and time are elements of consciousness. In a perpetual present, sensory impressions are identified with known memory contents and experienced in space and time. This is how we recognize the world (Bräuer 2005).

Man recognizes himself and the world, but only in part. A holistic perception of reality is not possible for self-aware people. The incomplete perception is the cause of illusion, conflicts and suffering, states Bräuer (2005). At this point the question suggests itself; how does evil come into our world? Which we want to answer following Bräuer (2005).

Selective perception is based on individuality. Space and time are aspects of our individuality. Space and time are not absolute and not independent of our consciousness. Space is consciousness space and time is consciousness time. We are not born into world space, this world space evolves with our individuality and thus our consciousness.

We basically hold: The natural sciences do not offer a world view, but only a view of nature, cf. on this Lesch (2016) in Chap. 2. That is, what the world is with its experiences.

With increasing individuality, our energy needs also increase. Obviously, being bound to space, time and matter is a problem for mankind. With our urge to expand our knowledge, to close gaps in our knowledge, we try to "get a grip" on this problem. But in order to do so, we again increase our energy demand. These considerations lead us to a world view, which we describe in Sect. 3.9.

Nature has let all living beings come into being in such a way that they, including us humans, can sufficiently cope in the world or in the spatiotemporal section of the four-dimensional spatiotemporal picture (i.e. in such a way that the further existence of the respective kind of living being is not endangered, otherwise the corresponding living being would have died out). Due to the high complexity of our brain, we are – only in a joint effort with other thinkers and experimenters – able to go beyond the view provided by the interaction of our sensory organs and the individual experiences in our lives, to develop a more general, and thus also a more inconceivable, view of the world as presented to us by the four-dimensional spatio-temporal picture, Bubb (1986), private communication.

Perceiving is becoming aware of something given in reality – through perception it becomes reality for us. We perceive the world in a certain way, which is not fixed, but is related to the structure of our consciousness. People of a culture share a view of the world which is expressed in a certain, elementary conception of reality.

The prerequisite for the process of perception is matter. For perception is based on memory, and memory is based on permanence, on unchangeability. And these are precisely the properties of matter. It is inert, it persists in its state. And thus it is the basis of memory and of consciousness. Consciousness is not possible without matter. Even if we do not yet understand the functioning of our brain by far, it is still matter in its highest form of organisation. Conversely, however, matter is inconceivable without consciousness. Consciousness and matter are thus interconnected (Bräuer 2005).

The framework of our perception also determines how we behave towards the world – it determines our patterns of reaction and action.

For the phenomenon of consciousness there are several schools of thought or different explanatory models. We have already introduced one with the two basic activities (negentropy ↔ information, Fig. 3.3).

In the explanatory model of cognitive psychology, consciousness is examined primarily in connection with information processing. The distinction between understanding (unconscious-automatic), perceiving (observing) and consciousness (consciously controlled evaluation, processing and storage processes) plays an important role here. Consciousness is closely related to the concept of attention and is understood as an orientation and control instance in processes of stimulus processing up to behavioral execution.

Fröhlich (2005) describes consciousness as "a hypothetical construct, an instance of information processing comparable to short-term or working memory, which is notable for its limited working capacity."

Perspective holds our consciousness in the past. Reality and consciousness are connected in a way that is not according to the cause-effect structure (i.e. consciousness as the cause of reality or vice versa), but they originally belong together and form a wholeness.

Our time consciousness plays a key role for the integrative performances of consciousness and their explanation from neuronal foundations, because what we experience as the present is, after all, identical with our contents of consciousness. What we focus our attention on is present; what we store in our memory is past; what our plans and intentions are directed towards is future, see here Chap. 2 (Falkenburg 2012).

With this we have presented our understanding of spatiotemporal behavior, which underlies our further considerations. Spatiotemporal behavior differs from mental (specifically: intentional) behavior. For the intentional structure we have used the Rubicon model. In Sect. 3.5 we will deal with the dualism of consciousness and intentionality.

3.5   Consciousness and Intentionality

Aristotle already described action as an intentional process. He already distinguished the intention of action from the anticipated goals of action and the circumstances (conditions) of action, cf. Chap. 2 (Hacker and Richter 2006).

Intentions cannot be explained scientifically at present. At this point, we again refer to the pre-scientific character of this paper, i.e., we cannot know what future findings will lead to. Conversely, intentions are paradoxically causally relevant, but it is not clear how they can be captured as causally relevant factors in the causal modelling of empirical conditional structures.

So when we talk about intentionality, we are discussing something that in cognitive science is seen as the intake of information from the environment and the behavioral dynamics that this triggers. In this context, the concept of information acts as a bridging concept. It closes the causal gap between physical and mental phenomena, i.e. between neuronal events and cognitive performances, such as perception, learning or remembering, or between brain and consciousness, per analogiam. For this, we rely largely on Searle (2006) in Chap. 2. He says that a huge evolutionary advantage of human consciousness is that we can simultaneously coordinate a large amount of intentionality ("information") in a single unified field of consciousness. Let's imagine the amount of coordinated information processing when we are driving a car. Let's not just think of the coordination of perception and action (for example, when approaching a red light.) Let's also think of the constant access to unconscious intentionality (digressive thoughts while driving, for example, about last night's dinner). All of these are intentionalist representations of the world that we use to master the world. Actions are performed out of intentions (reasons, purposes), regardless of what other conscious or unconscious mechanisms are involved.

Cognitive performances take place by means of biological processes, but they cannot therefore be identified with them, see Chap. 2 (Struma 2006).

Fundamental to the description of intentional processes is the identification of goals, their hierarchical organization, and the necessary feedback loops. Since the mid-1960s, related concepts of psychological theories of action have developed in mutual stimulation, which have the description of behavioral dynamics in common. Hacker and Richter (2006), cf. Chap. 2, have distinguished three "levels" of execution regulation of actions for the process of action regulation, the intellectual (novel, complex plans), the knowledge-based or perceptual-conceptual (flexible patterns of action), and the sensorimotor (automatic control), each of which can be subdivided again and therefore offer a relatively arbitrary number of descriptions.

Rasmussen (1986), see Chap. 2, has also presented three levels of cognitive behaviour (see Sect. 2.7, especially Fig. 2.2). They provide a good basis for identifying operational errors that are related to personnel requirements. Since not only errors were classified within the error taxonomy (classification), but also actions were already included with the thematically related accident origination model, a differentiation into unintentional and intentional actions was additionally introduced.

Searle (2006), cf. Chap. 2, developed his principle of individual and collective intentionality from the philosophy of language, addressing behavior that he also labels "rule-based," but with a different understanding than Rasmussen's.

Only Searle knows only "rule-based" or "rule-guided" behavior. For this he made an important distinction between the "prior intention" (intentional) and the "intention in acting" (non-intentional). The notion of error can be applied to intentional actions, see also Chap. 2 (Searle 2006).

It should be noted that, despite different terms for the same action regulation, the three models of action regulation presented stem from a common understanding of the concept of action.

After that, the delimitation of actions takes place through the conscious goal, which represents the anticipation of the result (anticipation) linked with the intention of realization (intention). Every action also includes cognitive processes via the goals.

Hacker and Richter (2006), see Chap. 2, say, "Goals are linkages of at least cognitive anticipation and motivational or volitional anticipation (of intention) and memory storage of anticipation as the basis of feedback target-actual comparisons."

The three theories also have in common that "levels" for the regulation of action are to be distinguished.

Rasmussen speaks of "skill-based" behaviour, where Hacker and Richter speak of "sensorimotor or automatic regulation". Rasmussen speaks of "rule-based" where Hacker and Richter speak of knowledge-based regulation. Rasmussen uses the term "knowledgebased" where Hacker and Richter speak of "intellectual" regulation. Hacker and Richter refer to the middle level in their model as "knowledge-based" because knowledge is converted into action relatively directly, i.e. without in-depth analysis, according to the stored IF-THEN rule. For Rasmussen, the "IF-THEN behavior" is also present at the middle level, but refers to this level as the "rule-based level". In Hacker and Richter's model, deep intellectual analysis is used at the "intellectual" level. We explain this difference in terminology by the fact that knowledge for the analysis and internal representation of the situation is understood differently in the two models.

Rasmussen as well as Hacker and Richter have in common the division into three main levels, as well as the assignment of the degree of consciousness.

The three levels of action presented here also lend themselves to a basic division along the lines of non-conscious, conscious but non-conscious, and conscious regulation.

For a better understanding, we would like to present the "levels" of action regulation in tabular form and, beforehand, once again describe our understanding of the three levels of action because of their importance using an example from everyday life.

For example, when we walk from our home to the train station, we can perform many of the processes that are necessarily involved "unconsciously" because they are skill-based and have become a daily routine. If we meet an acquaintance on our way, this meeting penetrates our consciousness, and we will greet the acquaintance – rule-based. If he also stops and wants to enter into a conversation with us, this happens at least partially on the knowledge-based level.

We have also included this example for two further reasons. Firstly, this example shows that the spatiotemporal framework of action can change unexpectedly through the addition of the known, and that other, further levels of cognitive regulation of action are thereby addressed. Secondly, we can see from this how time pressure can arise from unexpected events, assuming that we wanted to achieve a certain move.

There is also another aspect which is very obvious, namely the butterfly effect mentioned in Sect. 2.6.1.7.

Let's imagine that we leave our apartment just 1 min late, causing us to miss the train. We wait for the next train, but it is again late, causing us to miss the scheduled job interview assumed in this example, with the result that we don't get the open position we applied for. We have experienced the butterfly effect first hand, or in other words, experienced the effects of chaos theory. Such a concatenation of events is what mathematicians call "nonlinear phenomena".

If we had left our apartment just 1 min earlier, we would have, we assume in our example, gotten the open advertised job. A delay of just 1 min causes a chain reaction that leads to an unintended, negative result for us. Chaos theory, or the butterfly effect, states that minimal changes in initial conditions can have large effects on the entire system. To call such a constellation a coincidence would certainly be too short-sighted and would evade an answer to the question of direction.

This is referred to as unpredictability, which means that weather forecasts, for example, are only partially reliable because too many factors have to be taken into account.

Or in mathematical terms: Nonlinear equations can be found in very many scientific fields, such as astronomy and quantum mechanics, biology and medicine, or even the social sciences. The field of social sciences is particularly exciting for us and will be dealt with in Chap. 4.

Generally speaking, chaos theory, i.e. the butterfly effect, is one of the great scientific advances of the twentieth century, along with relativity and quantum mechanics.

Lesch (2016) in Chap. 2 addresses the same phenomenon, but considers it under the aspect of probability and calls it "tunnel effect". We quote: "Quantum mechanically, you have a vanishing probability that you "tunnel" through the wall". We stay with Lesch: "But you can't run into the wall that many times, as you would have to increase the probability, if you could do it even once within the lifetime of the universe." (Lesch 2016).

Table 3.1 "Levels" of mental regulation of action

Theories of action regulation				Rasmussen (1986)			Searle (2006)
Hacker and Richter (2006)							
Level of mental regulation		Analysis and internal representation of the situation (conditions of activity) (IF)	Action programmes (THEN)	Levels cognitive demand	Analysis and internal representation of situation (condition of activity) (THEN)	Action programmes (THEN)	
Consciousness	Levels						
Subject to consciousness	Intellectual regulation	Intellectual analysis	Strategies, plans	Knowledge-based	Novel situation	Conscious analytic processes and stored knowledge	Describes the conscious following of a causal rule as rule-following behaviour. In order for the rule to guide behaviour, the agent must be able to follow it voluntarily
Subject to consciousness but not required consciousness	Knowledge-based regulation	Perception of situation features/recall of (explicit) knowledge	Action schemes	Rule-based	Familiar problems, solution by means of stored rules	Conscious behaviour known situations	
Unable to consciousness	Automated automatic (sensori-motor) regulation	Picking up kinaesthetic signals/providing implicit knowledge	Automated automatic (motorized) programs	Skill-based	Stored patterns from pre-programmed instructions	Analog structures In spatiotemporal functional domain	

Lesch believes in "a grand cosmic relationship between us and the nuclei in the stars, the atomic nuclei." Lesch therefore answers the question of direction as follows: "Then it is no longer just the Very Smallest that is responsible for my existence, but the very smallest in the Very Largest." (Lesch 2016).

We have taken a deductive approach to the question of the direction of the three individual catastrophic events. We have also started with the Very Smallest, the scientific principles of thermodynamics and the theories of relativity, and related both to a sociological perspective which we have formed from the intentional structure of action and the spatiotemporal regulations of action.

For a better understanding of the social science perspective, let us take a closer look at the "levels" of mental action regulation, cf. Table 3.1.

We must say at the outset that the notion of unconsciousness is one of the most convoluted and ill-conceived notions in modern intellectual life, cf. Chap. 2 (Searle 2006). In general, unconscious reasoning is thought to be a basis of perception, along with sensations. The lower level consists of understanding (unconscious-automatic) or reflexive as well as skill-(completion-)based behavior. Unconscious actions occur reflexively. They arise from long practice when a particular action is required repeatedly in response to the same stimulus configuration. In a highly practiced state and under stable conditions of execution, the cognitive preparation of actions, but not the motivation and goal formation, is shortened to the retrieval from memory of ready programs that have become routine, see Chap. 2 (Hacker and Richter 2006). A special role is played by behaviour required by an order or instruction, which displaces motivation and goal formation from the action.

In the case of an order or instruction, the behaviour must conform to them, i.e. free will must be eliminated. They determine the action or, in other words, the behaviour must adapt itself in order to fulfil the content of the order or instruction. In an exaggerated form it is present in the military order. Compliance with orders, instructions and commands is punishable.

Compliance with an order or instruction was present in Chernobyl ("par ordre du mufti" on the part of the load distributor, "dispatcher"). The threat of punishment was the fulfillment or rather non-fulfillment of the five-year plan.

In Fukushima Daiichi, the punishable act was Japanese leadership culture.

On the Deepwater Horizon rig, the enormous cost and schedule pressures can be seen as punitive.

In all three cases, the constraints specified for the actions were understood and predominantly carried out at the conscious-but-not-conscious (Hacker and Richter (2006) in Chap. 2) or rule-based (Rasmussen (1986) in Chap. 2) level. This conceptual classification was chosen in order not to be misunderstood by the different conceptual world mentioned above.

Only this explanation ultimately remains for what happened in the three cases mentioned.

This behavior based on human understanding is to be distinguished from conscious behavior, with which the other two levels of cognitive information processing presented by Rasmussen and by Hacker and Richter (rule-based and knowledge-based or knowledge-based and intellectual regulation) are concerned.

It is aggravating to keep up with these different conceptual worlds. We therefore take a highly pragmatic step. We rely on Rasmussen's terms because, as already emphasized, they provide a good basis for identifying operational errors that are related to personnel requirements.

Behavior at all three levels is determined by perception and observation, as well as an increase in awareness of the processes. The regulation of action is determined for the person by following rules that causally determine his behavior. The rules for action are formed through analysis and internal representation of the situation at hand. In rule-based behavior, familiar problems are solved using stored rules. In knowledge-based behavior, novel situations are solved using conscious analytical processes and stored knowledge. The rule to be applied has a causal effect in producing the very behavior that counts as following the rule, see Chap. 2 (Searle 2006). If we take, for example, the rule of "driving a car to the right," then the content of that rule must play a causal role in bringing about our behavior. We think, for example, of driving a car. Not just the everyday routine one that is skill-(completion-) based, but especially rule-based driving. This doesn't mean that behaviour is entirely determined by the rule of "driving on the right", see overtaking for example. Nobody drives off just to follow this rule, but to reach a goal (intention). We have already addressed the necessary coordination in connection with the gaps in knowledge that occur in this process (see Sect. 2.10), which we will address from a further aspect.

Behaviour must therefore also have an intentional content that determines a certain aspect of the design. Rules are usually followed voluntarily at the conscious and conscious-obliging level. Rule-based and knowledge-based behavior is present when the agent can follow the rule voluntarily. More generally, we interpret the rule as enabling us to do things that are not determined by the content of the rule.

Society determines what is recognized as conformity with the rule. At this point, reference should be made to the catastrophic events described.

Behavior at the skill-based level takes place in a fraction of a second to a few seconds. Human behavior at the rule- and knowledge-based level takes place in a few seconds to minutes or a few minutes to hours or days.

Our ability to react is fundamentally independent of consciousness. Space and time have no meaning. Many things even work without consciousness, or at least they work better then. Reacting consciously to dangerous situations would sometimes be fatal. We have to act faster than we can experience it. We only notice the unconscious when it contradicts the conscious. Consciousness likes to know itself identical with the whole person and only reluctantly gives room to unconscious reflexes. Man is really happy when consciousness does not interfere. He feels most comfortable when he only acts. The consequence is: when we feel good, consciousness does not rule (Bräuer 2005).

With rule- and knowledge-based behavior, free will is a reality.

Whether a person has acted for the reasons he or she assumes cannot be decided with certainty either by the person himself or by anyone else, cf. Chap. 2 (Struma 2006).

We can neither derive nor understand consciousness from physical laws, although consciousness is fully compatible with the laws of physics. According to our understanding, consciousness is a property of the brain and thus part of the physical world. We thus distinguish ourselves from other definitions, especially the neurobiological one, which assumes that neurobiological experiments can solve the scientific problem of consciousness.

In short, there are gaps in rule- and knowledge-based behavior as a result of volitional awareness. Following a rule or scientific knowledge is characterized by the fact that one can either follow them or break them. Regenerating or even generating decision-relevant knowledge, the basis for knowledge-based behavior, takes time, which was not available in all three events.

In other words: Time pressure leads to the aforementioned gaps in knowledge, which we will deal with in the following.

The momentum of the time pressure seems important to us for a deeper understanding of the three individual events. Therefore, we will consider this momentum from a physical point of view.

The tacit convention resulting from the psychologically adapted sequence of before and after makes a sequence seem logical to us only if it is chronological, i.e. to the extent that the time direction corresponds to the increasing entropy. Thus, without really being aware of it, we have linked chronology and causality. We quote Grünbaum (1973) on this point: "… the convention that the direction of time is defined by increasing entropy is inseparable from the acceptance of causality as a method of explanation." We conclude that causal explanation depends directly on our adaptive sense of time. The maintenance of an open system (living cell or human society) amounts to slowing down the increase of entropy in the specific system, i.e., in the view advocated here, to time pressure. By creating information, organizing work, structuring actions, balancing the wear and tear of machines and concentrated energy use, we try to stop time by exposing ourselves to time pressure. In this respect man resembles a double-headed Janus. In him two different perceptions of the direction of time intersect, on the one hand he wants to compensate for the passage of time, on the other hand he wants to create a time reserve for himself. He wants to accomplish both through his creative activity, with which he creates potential energy or potential time. Potential energy and time are information. The creation of information (of potential time) takes place in the development of mankind in ever accelerating rates, we think here of the war of time, for example, through stock sales by computer. Negentropy, the objective measure of information, is, as we have noted, necessarily directed towards entropic time (cf. Fig. 3.3).

Summary

If one wants to achieve goals at a certain time, one must make a decision between various constraints. But every decision is necessarily based on a hierarchy of values in our society. It is inevitable that we will encounter gaps in our knowledge. We will deal with them in a moment.

3.6 Spatiotemporal Gaps in Knowledge

We have seen in Chap. 2, Sect. 2.8 that the connection between cause and effect or causality proves to be ambiguous in the analysis of the three individual events. This statement applies not only to the three individual events, but in general. In the field of natural history, we summarize this phenomenon under evolutionary epistemology, which states that organisms (including us humans) have "adapted" to the conditions of the environment and thus helped to ensure reproductive success in the face of all obstacles, so our "epistemological apparatus" has been shaped step by step in adaptation to the real world, see Chap. 2 (Penzlin 2014). By expanding our knowledge – that is, by closing gaps in our knowledge – we have succeeded in acquiring an ever more perfect and comprehensive knowledge of the world. So far, we have only made steady progress along this path, but not reached an end point. The role of knowledge gaps as a motor for progress remains.

We also revealed gaps in our knowledge of the three individual events due to the coincidence of the cause-effect structure and the intentional structure, which led to catastrophic results.

Gaps in cognition exist in both the cause-effect structure and the intentional structure. If there are gaps in causal laws of nature and in the cognitive model of action we use, then our free will – however it may act into the physical and mental world – can choose gaps in causality in order to activate precisely those causal chains in natural events or in the cognitive realm that we in and of themselves did not want to achieve with our actions.

Or, to put it another way, in analyzing the three individual events we found that there were inherent (inherent) gaps in knowledge in the technical system, nuclear power plant (Chernobyl, Fukushima Daichii) and in the Deepwater Horizon drilling platform, that were "dormant" unrecognized gaps in knowledge, which were revealed by intentional structures with which laws of nature were pushed aside or even ignored. "Slumbering" unrecognized gaps in knowledge were in Chernobyl the incomplete knowledge of the neutron-physical and thermohydraulic behavior of the reactor core, in Fukushima Daichii in particular the unprotected state of the emergency power supply, on the drilling platform the wrong drilling technique. The intended structures in the three cases can be stated in the same order of events with fulfilment of the five-year plan, economic operation of the power plant and compensation of the enormous cost and deadline pressure. In short, the "dormant" knowledge gaps were evoked by intentional structures that ignored the laws of nature.

Again, put differently or more succinctly: By the consequent reasons for action the exceeding of the point of no return was not recognized.

We have already referred to the time dependence of the point of no return (Sect. 2.6.1.5) in connection with column driving in road traffic.

We also find this time dependence in the three individual catastrophic events, but with different accentuation.

In the case of Chernobyl, the test was postponed by half a day by order of the Ukrainian load dispatcher. In order to ensure the continuation of the test, the last remaining safety system was switched off and thus the point of no return was set depending on the time arrangement from the load dispatcher.

In Fukushima Daichii, the point of no return was already formed at the beginning of power plant operation due to the insufficient (capacity) and tsunami unprotected independent emergency power supply.

On the Deepwater Horizon drilling platform, too much retardant was added to the cement, so that the mixture was still liquid after 24 h. But after 15 h, the drilling crew started to replace the drilling fluid above the cement with seawater. So the drilling crew set the early point of no return.

Why the crossing of the point of no return was not recognized by the operators remains the intriguing question. In Chap. 4 we want to answer this question by turning again to the frequently used phrase "par ordre du mufti".

We have described in the previous sections of this Chap. 3 how gaps in knowledge can arise through the spatiotemporal framework of relationships. Mainly by the fact that differently acting persons of the different hierarchy levels compete with each other in the spatiotemporal action frame.

Now we want to pursue the question whether "slumbering" gaps in knowledge can also become evident through the coincidence of the spatiotemporal framework of relations with the intentional structure.

Recall Fig. 2.3 and the Deepwater Horizon causal chain in Chap. 2. Assume that each link in the cause-effect structure and cognitive sequence of actions according to Heckhausen (1987, Quoted from Rasmussen 1986), cf. Chap. 2 is satisfied by consciousness and that each step has the particular kind of consciousness that reveals a gap, i.e., volitional consciousness. If we continue to assume, as we have so far, that the brain is non-deterministic, then we must ask, is there anything in nature that even begins to suggest the possibility of a non-deterministic system? Only the quantum mechanical part of nature contains a non-deterministic component. Predictions made at the quantum level are statistical because they contain random elements. Quantum theory states that only the probability of a process can be described.

We have illustrated the connections between classical mechanics, special and general relativity with Fig. 3.1. General relativity is a reference system with curved light paths and corresponds to a gravitational field, vice versa in a gravitational field the light paths are curved. This can also be measured. Since the experimental proof in 1919/1921 we know that masses bend space. Not only space, but also light travels along the so-called spatiotemporal. In the theory of relativity, we speak of projections of our current world view onto a time when this world view and this kind of space and time did not even exist. Quantum mechanics also points in this direction. Quantum mechanics describes the temporal development of matter.

In physics, we refer to quantitative, calculable relationships. These quantitative relationships are based on our objective observations. We break down the world into observable details and place them in spatial and temporal relationships to each other, which are calculated with the help of coordinate systems.

This computability is a very special aspect of our individual experience of the world. Many other aspects, especially somewhat more complex systems, are not computable, but show the same structures. And thus the process of becoming explicit of contents of consciousness becomes directly experienceable for each of us (Bräuer 2005). But it is also known that not everything is subject to direct observation, e.g. consciousness. So we are subject to limitations related to our perception of time. It follows that "the convention defining the direction of time by increasing entropy is inseparable from the acceptance of causality as an explanatory model" (Grünbaum 1973).

De Rosnay (1997) also addresses the question of how consciousness and the universe "interlock" in a dialectical process of observation and action. He thus integrates the realities of thermodynamics, information theory, and the physics of relativity. He goes on to state that entropy is the measure of the lack of information about a system. As we have already noted, information is also the equivalent of negative entropy. Every experience, every action, every acquisition of information by the brain consumes negative entropy. You have to pay a fee to the universe: irreversible entropy.

John R. Searle says:

> Arguably, the evolutionary function of consciousness may be, at least in part, to organize the brain so that conscious decision-making can occur even in the absence of causally sufficient conditions, even if the effect of conscious rationality is precisely to avoid making random decisions, see Chap. 2. (Searle 2006)

Go on there:

> All that is being said is that, given our current state of knowledge, the only established non-deterministic element in nature is the quantum level. Therefore, if we are to assume that consciousness is non-deterministic, that the gap is real not only psychologically but also neurobiologically, then given the current state of physics and neurobiology, we must assume that there is a quantum mechanical component in the explanation of consciousness. (Searle 2006)

Selected behavioral aspects will lead to a further explanation of consciousness by deepening the differences between the three levels of cognitive regulation of action.

3.7   Selected Behavioural Aspects

We will deal only with those behavioral aspects that can be purposeful for an answer to the question of the reliability of human action or direction.

In Chap. 2, in Sect. 2.3, we introduced the cognitive science approach to information processing, in particular Jens Rasmussen's model of mental information processing in terms of the degree of cognitive demand on people. In Sect. 2.5 we looked at Heckhausen's holistic cognitive model of individual action.

We would like to say in advance that the term cognition is a collective name for all processes or structures related to awareness and cognition, such as perception, memory (recognition), imagination, concept, thought, but also assumption, expectation, plan and problem solving.

Cognitive performances, as said before, are carried out by biological processes, but they cannot therefore be identified with them, cf. Chap. 2 (Struma 2006).

With this understanding we can continue to speak of the three cognitive levels of behaviour according to Rasmussen (1986), see also Chap. 2, which we want to extend to include spatiotemporal behaviour.

We have already extended Rasmussen's three levels of behavior by defining under automated/automatic (sensorimotor) behavior (Hacker and Richter) or under ability-(skill-)based behavior (Rasmussen) an unconscious, reflexive behavior, under knowledge-based behavior (Hacker and Richter) or under rule-based behavior (Rasmussen) a behavior that is capable of consciousness but not subject to consciousness, and under intellectual behavior (Hacker and Richter) or knowledge-based behavior (Rasmussen) a behavior that is subject to consciousness after elaboration of the reasons for the decision (need for time). We would like to extend this understanding by further aspects of spatiotemporal behaviour, namely understanding, awareness (perception and observation) and consciousness (Bräuer 2005).

We have already mentioned understanding in connection with unconscious actions. Here, understanding is not understood acoustically, but, for example, as the fulfilment of an instruction given to us. Such an understanding was present in the case of Chernobyl; here the instruction of the load distributor was "understood"; this "understanding", however, did not correspond to the "understanding" of the physical connections which we must presuppose with the operators – keyword controlled neutron population.

Bräuer (2005) distinguishes between understanding and perception. He understands this to mean the perception of a process, e.g. sunset. However, this observation is not connected with associated time sequences and/or physical laws. It is an intense experience of the world, solely in the present. Neither past nor future play a role. Time does not exist. The human being registers an event without an evaluation for possible consequences of action. He does not think about the associated time processes and physical laws. Space has no meaning in perception. Thoughts and feelings arise, this is the aspect of perception.

Consciousness is a subjective matter and does not necessarily require rigorous logical cognitive reasoning.

Or in general terms: Consciousness develops in the confrontation with everyday problems and the major problems of the world (Bräuer 2005). It is always about the struggle for survival, in a biological and/or social sense. We are constantly in some kind of conflict and grappling with it. "Without conflict, thoughts rest, and then consciousness transitions into awareness. Conflict is also always linked to suffering" (Plank et al. 2012), see Opening Pandora's Box. People experience themselves as individuals and have to face the struggle for survival and solve the ever-changing

Table 3.2 Merging the behavioural levels under the aspect of spatiotemporal behaviour

Cognitive behavioural levels according to Rasmussen		
Skill-based behaviour	Rule-based behaviour	Knowledge-based behaviour
Cognitive behavioural levels involving spatiotemporal behaviour		
Understand	Awareness	Consciousness
Reflexively unconsciously	Registration of events without evaluation for a possible action	Conscious application of knowledge-based level
Timing of implementation		
Instantaneous, immediate, fractions of a second to a few seconds	Few seconds to minutes	Time required to activate knowledge resources

Note: Varying between the different behavioural characteristics is characteristic of all types of behaviour and occurs in dependence on the respective perceptions in the real situation on a rule- and knowledge-based level according to free-will decisions

problems. In the process, consciousness and corporate culture develop, which interests us in the context of the three individual events.

In short: space and time are elements of our consciousness (Bräuer 2005).

Related to time: We deal with time as if it were space, entirely according to our inner, conscious experience of time. Time is the conscious experience of change. Because individual contents of consciousness are the same in all spatial, temporal and causal relationships, these can be grasped mathematically. We confuse the mathematical relations with reality and thus make it impossible for us to comprehend relativity of space and time (Bräuer 2005).

The variation mentioned in Table 3.2 is based on volitional decisions, for which there are various approaches, which we have again selected under the aspect of the reliability of human action.

Determinism denies freedom of will in general, indeterminism claims that the irrational "core of the person" makes the ultimately moral decisions, even if the will is otherwise largely determined by "external factors". Freedom of will is presupposed for ethically attributable action (e.g. partly in criminal law). Psychologically, the decision of will is interesting because the fact that the observation of experience gives the impression that one could have acted differently than one has acted.

Esfeld (2017) also says, "Physical determinism says nothing about free will." His observation, "Under stable environmental conditions, deterministic laws could also be conceived in biology and even in psychology and the social sciences." This observation, applied to goal-directed (teleonomic) processes (consequential reasons for action), means that they are – first – controlled by programs and – second – have

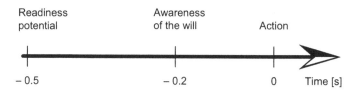

Fig. 3.4 Conscious volition is experienced one third of a second after the readiness potential! This figure has in (Bräuer 2005) the no. 8–3 and the same title

a goal point at which the process is terminated. According to Penzlin (2014), cf. Chap. 2, internal programs that are genetically fixed are responsible for goal-directed processes in nature. Just as the thermostat does not determine the temperature of the home, it is the human who sets the "set point". If this "set point" leads to conflict with the feeling of well-being in the home, it is not the "set point" that is responsible, but the person who has set the "set point".

Or, adopting Esfeld's (2017) conclusion, "Nature does not provide us with the norms for shaping human life and coexistence. With that freedom, we have responsibility to set those norms ourselves."

So we have to put some kind of "constraint" on the interpretation of the laws of natural science in order to realize the consequences of our free will.

But this is not considered a decisive argument for free will. Therefore, one has tried to solve the question of free will with experimental middles.

Ever since Benjamin Libet (1985) conducted his epoch-making experiments in the early 1980s in two series of experiments on free will, the results of which provided an astonishing picture, there has been a bitter dispute about free will.

With his experimental results Libet shows the complexity of the processes of consciousness. In the time of half a second to a whole second brain signals can be disturbed and thus conscious impressions can be prevented.

Our actions start unconsciously. Even when we think we are consciously deciding to do something, the brain is already active half a second before the decision. Not consciousness, but unconscious processes are at the beginning (Bräuer 2005).

Consciousness makes us believe that it makes decisions and is the author of what we do. But when the decisions are made, it is not present itself. It lags behind time and makes sure we don't notice. Measurements, particularly those of Libet, attempt to understand consciousness as a materially anchored quantity based on activity in the brain. As Fig. 3.4 shows, consciousness cannot be primary; it is not at the beginning. Some activity must already be there before consciousness gets going. What is strange about this is that we experience the decision consciously only after a significant delay, after it has already been made. Since only the conscious is conscious, consciousness cannot determine (Bräuer 2005).

According to Bräuer (2005): "It is like the blind spot of the eye: our perception of the world is faulty, but we do not experience the faults. The consciousness adjusts itself with a delay and does everything to hide this fact – from itself. It deceives. It deceives itself. This is very expedient – if you have enough time."

Benjamin Libet's exploration of consciousness processes and their relationship to the brain comprise two complexes. First, the realization that an activity of half a second's duration is necessary in the brain before something enters consciousness. And at the same time that consciousness makes a subjective temporal reference back. Second: that the consciousness of a decision to perform a certain action occurs about 0.35 s after the moment when the brain became active.

In summary, both statements provide the following picture:

About half a second of brain activity must have taken place before consciousness arises. This is true for sensory perceptions as well as for decisions. In the case of perceptions, however, the experience is backdated in time so that it is felt as if it occurred at the time of sensory stimulation. In the case of conscious decisions to act, the conscious decision is experienced as the first link in the process, while the brain activity that began barely half a second earlier does not enter consciousness.

Consciousness presents man with an image of the world and an image of himself as an acting subject in this world. Both images are highly reworked. The image of sensory perception is, insofar as parts of the organism have already been influenced by it for up to half a second before consciousness learns of it. It hides the subliminal (subliminal) perception that may be present – and reaction to it.

Consciousness is not a top-level entity issuing instructions to subordinate units in the brain, but a selective factor that makes choices from among the many possibilities offered by the non-conscious. Consciousness functions by sorting out suggestions and confusing resolutions proposed by the non-conscious. It is segregated information, cashed-in possibility (Bräuer 2005).

Libet (1985) further developed his experimental results around the veto principle. He conjectured that while consciousness cannot initiate the action, it can decide that it will not be realized. That the veto process is usually associated with discomfort does not mean that we cannot exercise such a conscious veto. The veto is there, even if we don't bring it to bear. The veto principle states that the veto itself is not unconsciously initiated, but occurs directly at a conscious level. However, he did not base this assumption on experimental findings (assumption!), but instead pointed out that alternative conclusions had led him to it. The veto principle, he said, is familiar to us from the history of human ethics. Referring to "many ethical constraints in particular the Ten Commandments, which are instructions on how not to act," Libet (1985) writes. Recent experiments on the consciousness of volitional decisions can be understood to suggest that veto decisions are also made unconsciously and only subsequently perceived as free choices. Libet's original and more far-reaching interpretation of his results would thus have been subsequently confirmed decades later. Libet Experiment: https://de.wikipedia-org/wiki/Libet-Experiment.

The veto principle has become part of human ethics.

We clarify once again: Whether a person has acted for the reasons he assumes can neither be decided with certainty by himself nor by any other person.

From these sensory impressions and memory contents of the brain pointed out by Libet (1985), a coherent world view develops in our consciousness.

One aspect of this is the spatiotemporal image of our experienced world, which we would like to round off with a quotation from Bräuer:

> Space and time are aspects of our individuality. Space and time are not absolute and independent of our consciousness. Space is consciousness space and time is consciousness time. We are not born into world space, this world space evolves with our individuality and our consciousness of it.
>
> With our individuality we have developed an enormous need for energy … [which we have already addressed] …. Obviously we suffer from the bondage of our conscious being to space, time and matter. With our physical knowledge we try to alleviate this suffering. And the energy demand for this is constantly increasing. Perhaps a world view that encompasses more than the purely physical aspects of our existence will lead us to other possibilities of well-being. (Bräuer 2005)

Another aspect, besides the already mentioned individuality and the perception of our responsibility, also determines our actions.

The people before our time knew nothing about themselves, at least we have no record of them to give us any information about their thinking. They certainly had no conception of individuality, nor of a space that separated them from their environment and other tribes, nor of a time in which they were born and died, and in which danger, hunger, and pain awaited them. All the knowledge that is had of humans and precursor species of humans relates to additions in graves. Humans are obviously the only living beings who buried their dead and also thought about their finiteness and formed ideas about it. At some point, humanity developed the urge for self-knowledge and with it the urge to transcend boundaries. The central insight from this is that contained in the biblical account of the expulsion from paradise: By eating from the "tree of knowledge," man reflected on his being and was therefore expelled from paradise, Bubb (1986), private communication.

According to Hegel, every boundary challenges the attempt to transcend it. Man created a stage for himself in which he could experience himself. And the basis of this stage are space and time.

On this, man practices self-expression, emphasizing his urge for increasing individuality. Individuality has become a high and important value for him in today's Western society. Individualization is predominantly forced in a unity of script, direction and representation. The individual writes his own script, often using spontaneously formed opinions; opinions formed in this way are today circumscribed by the term post-factual (emotional), i.e. they are not backed up by scientific facts. Factors for directing are majority ability, resonance in the media, emphasis on individuality through egocentric advocacy of self-interest, and gaining sympathy in the group.

With his self-portrayal, man tries to bring the script and the direction into harmony. If he succeeds, his world view is characterized by satisfaction with his environment. If he does not succeed, his behavior is determined by self-doubt, which can lead to "evil".

In other words, centrifugal forces within society are increasing. Individuality increases self-worth and at the same time reduces social cohesion.

There is no common, accepted value system of the social and culturally connected community and thus no general consensus for the evaluation of the three individual events presented.

Bräuer (2005) explains the phenomenon of increasing individualization as follows: "People identified with the acceptable aspects of their spatiotemporal existence and they projected differently away from themselves, into nature and fellow human beings. And so, I suppose, evil came into the world as a projection of one's own archetypal shadow."

We adopt this, admittedly pessimistic, judgement because the egoism of "evil" plays a dominant role (very vividly illustrated in the role of Baron Scarpia, chief of police in the opera "Tosca", a musical drama by Giacomo Puccini) and all calls to push back this kind of egoism go unheard. We will deepen this approach in Chap. 4.

In addition to this lack of a consensual system of assessment supported by the majority of humanity, another aspect should be addressed that goes hand in hand with increasing individuality, namely the associated increase in energy requirements already mentioned.

The energy demand increases with increasing individuality, even dramatically.

In his inaugural lecture on November 10, 1986, Heiner Bubb (1986) presented the following example, which aimed to show that, entirely in the sense of the accelerated increase in entropy also observed in the cosmological realm, developmental processes in the personal environment also proceed in such a way that the increase in entropy is forced. "On a historical view, it is also actually to be noted that man tries to make himself independent of the contingencies of nature through energy consumption," he writes (Bubb 1986).

A comparison of the amount of energy required by a pedestrian and a passenger car to cover the distance of 100 km indirectly shows the gain of entropy increase. But it also shows – using the example of striving for higher travel speed – that our needs are obviously such that they can only be satisfied by energy consumption and the associated entropy increase. (Bubb 1986)

We continue to quote:

The increase in entropy is nevertheless to be expected, since the corresponding use will be made by more and more individuals and since new needs will also arise again and again, which can be satisfied by further energy consumption. (Bubb 1986), cf. also Fig. 3.5

The increase in entropy is also reinforced by the use of increasingly comfortable means of transport by more and more individuals, which thus leads to further energy consumption.

We would also like to see this entropy increase from the point of view of concentrating and channelling energy, i.e. stopping time, stopping it from getting lost.

Or, in other words, to stop the passage of time and to live in the intensity of the moment (we recall the quotation already used [Sect. 3.3]): "Will I say to the moment: Stay! You are so beautiful! Then I will gladly perish!", cf. to stop and not to strive towards finitude.

With this example, the link between the passage of time and energy can be seen. The law describing this connection is very simple: "Buying time or gaining time" is paid for with energy. You use a car to get to your destination faster and more

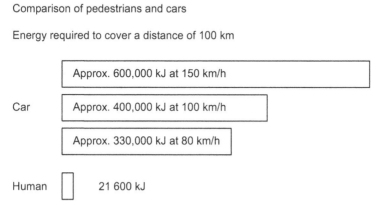

Comparison of pedestrians and cars

Energy required to cover a distance of 100 km

Car

Approx. 600,000 kJ at 150 km/h

Approx. 400,000 kJ at 100 km/h

Approx. 330,000 kJ at 80 km/h

Human   21 600 kJ

Fig. 3.5 Energy consumption of pedestrians and cars to travel a distance of 100 km; taken from (Bubb 1986). The figure is not to scale

comfortably, you use assembly lines and automation to be able to produce faster and to have more free time. The time gained has to be paid for with additional energy consumption. This amount of energy put into the whole social system must always be increased in order to be able to increase economic growth and supposedly gain time. It is the creative act that makes it possible to compensate for the passage of time, because every machine, every construction that makes it possible to compensate for the passage of time is equivalent to potential time.

Another point of view should not remain unmentioned here: Humans have increased their physical forces a hundredfold through technical energy transport and conversion. This has also increased the potential danger a hundredfold, and the responsibility of the developers and users of technical systems has increased accordingly. As long as this group of people remains aware of their increased responsibility and takes sufficient safety measures, their actions are justifiable. Only when a "casual" sense of responsibility prevails does their approach become a threat. The approach "the end justifies the means", i.e. the consequent reasons for action, as happened in the case of the three individual events, on the other hand, is not justifiable.

At what price?

Entropy increases, errors become more frequent. At some point, the "time reserve" of the system is exhausted; keyword aging management. The system has reached its finiteness.

In the Deepwater Horizon disaster, BP ended up having to pay US$45 billion in damages for the largest environmental disaster in U.S. history to date – roughly a thousand times what the company was trying to save by negligently shortcutting safety requirements and "buying time."

The cost of the reactor disaster at the Fukushima Daiichi nuclear power plant in Japan is reported at EUR 177 billion in 2016. Also more than a thousand times the expenses saved by not designing the emergency power system "according to requirements".

An undead man rests in the sarcophagus of Chernobyl. According to experts, up to 200 tonnes of uranium and plutonium are still lying dormant in reactor 4 of the destroyed nuclear power plant. The construction of the new shell alone cost EUR 1.5 billion, and follow-up costs are calculated at EUR 700 million.

Our epoch, which is so greedy to "buy time", i.e. to gain, is characterized by "buying time" as much as by the conquest of the universe. The decisions that have been or are being made in this regard are made on the premise that society recognizes only human beings and not nature, and all human beings as individuals, thus creating a collective memory.

More generally, time is the decisive factor in all questions of our social community. From a physical point of view, this concerns all perspectives of human existence, because time is quasi "imposed" on us by physical processes and as a subjective sensation.

3.8   Spatiotemporal Relationship Framework for the Three Individual Events

In Chap. 2 we have underlaid the heuristic developed there with the metaphor of archery by Mario Bunge (1959), see Chap. 2, in order to be able to clarify the links of the cause-effect structure.

In this section we want to proceed in a similar way. We introduce a metaphor for spatiotemporal action and, in doing so, develop parameters for spatiotemporal behaviour with which to assess the occurrence of the three individual events – the Chernobyl and Fukushima Daiichi reactor disasters and the explosion of the Deepwater Horizon oil rig.

The metaphor used is the use of public transport, here the public bus, which Witzleben (1997) adopted from H.-J. Engler in his dissertation "Handeln, Erkennen und Selbstbewusstsein bei Kant und Fichte" (Engler 1982). This metaphor is modified for the explanation of spatiotemporal behaviour.

If you want to use a bus, you choose the appropriate bus according to the deadline for reaching the destination – intention. With the calculated time for the walk from the place of stay we move to the bus stop. We map the bus at a certain point in time in our spatiotemporal action space, which is formed by our movement speed and the bus already on the route. If we see the bus already at some distance from us, we have to estimate the speed of the bus and thus its distance from the bus stop and adjust the speed of our movement so that when the bus arrives we reach the bus stop "just in time" and in the correct order, first the passenger, then the bus, "just in sequence". The bus driver will not approach the bus stop if he does not see a potential passenger or if no passenger wants to get off. The action space is therefore not yet fixed at the time of its opening, the beginning of the action. It depends on the mutual accessibility. This trivial case reveals further structures that make the variables of spatiotemporal behavior recognizable.

The fact that the bus is coming is initially an observation in our spatiotemporal space of action. We approach the bus stop and are sure that we will reach it in time before the bus if the speed of the bus is correctly assessed. However, this observation is at the same time an action-relevant event that depends on how the driver of the bus behaves. If he drives faster, this results in changes for our action parameters in order to ensure that we can reach the bus. Ensuring our intuition is thus dependent on the spatiotemporal behavior of third parties, in this case bus drivers. The metaphor shows on the one hand that our spatiotemporal observations fix our action space and on the other hand that it represents a coordination product that we can only partially influence.

More generally, not all relevant information is accessible to the observer, such as the current speed of the bus, unpredictable events such as a random meeting of a known person. This also applies to parameters that follow from the actions of other people, which cannot be anticipated. After all, it could be that the bus driver is waiting to pick us up as a passenger, or that he is passing the bus stop because the bus is crowded or he is desperate to keep to his schedule. We have two spaces of action here, the bus driver's and our own, both of which must fit together in order for our intention – to reach our chosen destination on time – to succeed. However, we can only influence the parameters of our spatiotemporal action space.

The situation is so complex because the space of our action can at the same time be the observable space of action of further persons who insert other elements of action into our spatiotemporal space of action that were not related to the initial space of action. However, the determination of the action space can only be done a posteriori, which makes analysis difficult. There are certain standards for analysis; we focus on reliability, which is done from a subjective perspective. This brings into play a quantity that has an ethical quality and that is constitutive of individual action. The ethical dimension requires that one's own goal is prioritized and that the goals of others follow this priority.

Thus, one's own goals are removed from doubt, and other people's goals become variables. Since this strategy can be followed by all participants, each of the participants will pursue his values, which he considers indispensable, and put aside other values, which are more open to him and therefore negotiable. However, this mutual value-forming process cannot really be predicted for real decision-making situations, just as it cannot be predicted for the borderline point of breaking off the decision-making situation, e.g. as a result of resignation. But the process of value formation is and remains a part of action.

Before we turn to the action spaces of the three individual events, we want to summarize the spatiotemporal action space of the "public bus".

The central area of action for the bus driver and the potential passenger is the timely accessibility of the bus stop, quasi the coordination point.

The bus driver's scope of action is defined by the route, the timetable and the passengers. We disregard the unpredictable.

The action space of the potential passenger is determined by his own speed and his observation, with regard to unpredictable influences we proceed in the same way.

In short: our actions are subject to the overlapping and acceptance of other bodies and persons influencing our spatiotemporal space of action.

The space of action can be conceived for the non-determinable "world-open" being of man only through his consciousness. His action becomes a spatiotemporal event that cannot be explained by causal relations in a deterministically presupposed space of action.

Or in physical terms: The spatiotemporal relational frame is a physical structure to be able to describe the behavior of location- and time-dependent bodies, which can also be influenced by communication-dependent quantities (persons). In the spatiotemporal relational frame, events are described by the position and movement of bodies under physical/communicative influences. The description also establishes the reference point. This is necessary because different observers describe the same event differently.

It is argued here that there is no clear way other than the cause-effect structure – consisting of the preliminary phase, generation of an internal state (cause), entry of an external system, point of no return, triggering event (causal principle), probabilistic influencing factors, effect – in order to be able to show and assess what is happening.

The assessment of variables in the spatiotemporal relational framework is and remains complex because it is shaped by our own intentions and thus our purposes, desires, goals of action, etc. and our own space of observation. Vice versa, this also applies to the spatiotemporal action space(s) of other social groupings or individuals by which our action space is or can be influenced.

Nevertheless, an attempt is made here to mirror the three individual events against the parameters of spatiotemporal action.

3.8.1 Chernobyl

The central area of action was the reactor core of unit 4 in Chernobyl (keyword: energy generation), both for the load dispatcher, whose task is the consumer-oriented distribution of electrical energy via the grid structure, and for the operator of unit 4, who is responsible for controlling the neutron population. The load dispatcher determined "par ordre du mufti" (our understanding of this term was given in Chap. 2) the experimental procedure in such a way that a consumer-oriented supply of electrical energy remained guaranteed. His space of action occupied the space of action of the block operator. He was deprived of it and thus it was not possible for the operator to ensure compliance with the laws of neutron physics (keyword: xenon poisoning).

From a spatiotemporal point of view, this is a complete occupation of the action space of the block operator by the load dispatcher.

Note: The meaning we attach to the expression "par ordre du mufti", which we use frequently, is explained in detail in Chap. 4.

3.8.2   Fukushima Daiichi

The first of the total of seven tsunamis arrived at the site 41 min after the seaquake. After the occurrence of the largest tidal wave of more than 14 m in height, 12 of a total of 13 emergency diesel generators failed (see Fig. 2.6, Chap. 2).

Obviously, the emergency power case due to external impacts was not investigated within the framework of a probabilistic safety analysis (PSA). (We have already addressed the incompleteness of the safety analysis in connection with possible hydrogen explosions). It can be safely assumed that 41 min was not sufficient to protect the emergency diesel generators against flooding. But the 41 min were certainly sufficient to ensure an alternative feed-in of electrical energy via reserve and external grids to be installed at short notice or not used. In units 5 and 6 of Fukushima Daiichi, at least one of the two emergency generators probably functioned during the earthquake and tsunami. This was sufficient to cool the two reactors sufficiently. Nevertheless: The saving electric current of block 5 and 6 was only a few hundred meters away. This might have been enough to supply at least the most necessary safety systems in units 1 to 4 (light in the control room and a feed pump for each unit, if necessary in rotation under the units) with electrical energy.

Why didn't they build a bus bar in the 41 min that would have routed electrical power from one generator to the other units?

As a further possibility for the power supply of the power plant, in case of emergency power failures due to disturbances caused by external events, a gas turbine with the possibility of cold start (without external energy), which is connected to the plant by a protected underground cable, is available. Obviously, at Fukushima Daiichi, this possibility of power supply independent of external events did not exist. (In Germany, this possibility is available).

Even when the site of the Fukushima Daiichi power plant was selected, the historical data for earthquakes and tsunamis had not been comprehensively evaluated. It would therefore be urgent to site such a gas turbine where the danger from tsunamis could be excluded on the basis of historical data, cf. also Fig. 3.6.

The fact that safety was not taken into account in the selection of the site for the plant and the design of the emergency power supply meant that the operators at Fukushima Daiichi had no room for manoeuvre. It is hard to say that the operators were completely at the mercy of the tsunami waves.

In short, there was no room for manoeuvre for the operators at Fukushima Daiichi.

Fig. 3.6 On the coast of Japan there are hundreds of these marker stones with the inscription: "Do not build below this stone" or "In case of earthquakes, watch out for tsunamis!". (Picture alliance/ ASSOCIATED PRESS)

3.8.3   Deepwater Horizon

We have already explained that barriers to safety involve loss of time, other inconveniences, and additional expense. Man is like a two-faced Janus. He is the place where two qualitatively different perceptions of the direction of time meet. Nevertheless, it is his individual creative act that enables him to compensate for the passage of time, as the operating crew on the Deepwater Horizon drilling platform aspired to do. In this striving, the operating crew restricted its actual scope of action to such an extent that it could no longer overlook the development of the explosive gas cloud in particular and thus "headed for" the oil catastrophe, supported by the regulatory authority.

3.8.4   US Airways Flight 1549

We cannot help but also cite a positive example of the use of the action space.

On 15 January 2009, an Airbus A-320 (US Airways Flight 1549) flies into the middle of a flock of geese in New York, the birds smash into the fuselage and get caught in the engines. Both fail immediately. The captain reaches for the stick and takes over the plane, saying, "My plane." The co-pilot nods in agreement: "Your plane". A clear assignment of the action space. What follows is 3 min and 28 s that will go down in aviation history. In the end, 155 people are saved in an emergency ditching in the middle of New York on the Hudson River. It was the best decision under the circumstances.

The pilots have different "ditching (landing on water) checklists" (Romero 2009). The pilot and co-pilot decided to use the shorter checklist because the pilot noticed that his co-pilot had to constantly switch between the displays on the screen and the manual while working through the checklists. To the pilot, combining the two systems seemed illogical. In 3 min and 28 s, which elapsed from bird impact to touchdown on the Hudson River, the co-pilot would not have been able to work through the checklist. Only an abbreviated checklist was applicable at the low altitude because the influencing parameters such as the size and direction of the waves, the altitude, the descent rate of the aircraft were not included in the checklists. How can a human make correct decisions in a time of 3 min and 28 s?

The pilot dispensed with all the information that the on-board computer had provided him with. He took the reins of action into his own hands, the only way he could make the right decisions at the right time.

Flight electronics is a double-edged sword. The systems are becoming ever more precise, and therefore also demand extremely fast and timely decisions. These are pushed into the background by the autopilot, but they are necessary for survival, as Flight 1549 shows us.

This tension shapes our view of the world.

3.9   Our World View

Our epoch, which is so greedy for gaining time, puts on fetters in its spatiotemporal actions through the conquest of time and space, which should actually be a social consensus to recognize.

We have computerized means of communication and transportation, and are straining to develop more and more machines to save time. Computers operate on a nanosecond scale (a nanosecond relates to a second as it does to 30 years). The conclusion is: the amount of information necessary for the functioning of highly complex processes exceeds our ability to process it, even if we are assisted by the computer.

We must get to the point, as we have established in the context of time pressure, of leaving operations determined by the goal of action to humans and not to be determined by machines. Only in this way is it possible for those responsible to make decisions, allocate time and resources and ultimately organise time sequences.

The completion of Flight 1549 was only possible because it was a destination-determined operation for the pilot and co-pilot. This allowed both of them to make the right choices, divide the short time remaining according to the phases of the flight, and miraculously complete the flight.

We would like to emphasize once again that we clearly define our goals of action and when we want to achieve it, and constantly review this chronicity during our action. If one wants to achieve goals at a certain time, one has to make a choice between different constraints. Every choice is necessarily based on a hierarchy of values and thus on free will.

We should hold on to that!

We have attempted to provide an explanation of our worldview that allows mental phenomena a proper place in the natural world, represented by the cause-effect structure and the spatiotemporal continuum. We have addressed mental phenomena in all their facets – consciousness, intentionality, free will, mental reasons for action, understanding, perception, observation, unconscious and conscious action, etc. – on the premise that they are part of the physical world.

Searle says: "We should understand consciousness and intentionality as being as much a part of the natural world as photosynthesis and digestion. ... because consciousness and other mental phenomena are biological phenomena. They arise from biological processes and are peculiar to certain kinds of biological organisms." cf. Chap. 2 (Searle 2006).

This is our view of a scientific worldview. In our view, science denotes methods of finding out about something that admits of systematic investigation. Our cognitive-scientific worldview is based on intentionality and consequentialism and thus on thought-law structures that are shaped by free will.

We come back to an earlier statement that we do not live in many worlds, nor do we live in two different worlds, a mental world and a spatiotemporal world, a scientific world and an everyday world. There is only one world in which we all live, and we must always keep it in mind in our actions.

Through his mental, spiritual ability – through his capacity for consciousness – man gains a creative power with which he must deal responsibly.

References

Bräuer K (2005) Gewahrsein, Bewusstsein und Physik: Eine populärwissenschaftliche Darstellung fachübergreifender Zusammenhänge. Logos, Berlin

Bubb H (1986) Energie, Entropie, Ergonomie, Eine arbeitswissenschaftliche Betrachtung. (Eichstätter Hochschulreden 60). Minerva Publikation, München

Bunge M (1959) Causality, the place of the causal principle in modern science. Harvard University Press, Cambridge

Carrier M (2009) Raum-Zeit. De Gruyter, Berlin

Costa de Beauregard O (1963) Le Second Principe de la science du temps. Éditions du Seuil, Paris

de Rosnay J (1997) Das Makroskop, Systemdenken als Werkzeug der Ökogesellschaft. rororo sachbuch, Reinbek

Der Brockhaus Naturwissenschaft und Technik (2003) F. A. Mannheim. Leipzig und Spektrum Akademischer Verlag, Heidelberg

Engler H-J (1982) Handeln, Erkennen und Selbstbewusstsein bei Kant und Fichte. In: Poser H (Hrsg) Philosophische Probleme der Handlungstheorie. Alber, Freiburg i. Br

Esfeld M (2017) Erkenntnistheorie Wissenschaft, Erkenntnis und ihre Grenzen. Spektrum der Wissenschaft 8.17. Springer, Heidelberg

Falkenburg B (2012) Mythos Determinismus. Wieviel erklärt uns die Hirnforschung? Springer Spektrum, Heidelberg

Fröhlich WD (2005) Wörterbuch Psychologie. Deutscher Taschenbuch Verlag, München

Grünbaum A (1973) Philosophical problems of space and time. Reidel, Dordrecht

Hacker W, Richter P (2006) Psychische Regulation von Arbeitstätigkeiten in Enzyklopädie der Psychologie, Themenbereich D Serie III Bd 2 Ingenieurpsychologie. Hogrefe, Göttingen

Hawking SW (1994) Eine kurze Geschichte der Zeit. Die Suche nach der Urkraft des Universums. Rowohlt, Reinbek

Heckhausen H (1987) Perspektiven des Willens. In Heckhausen H, Gollewitzer PM, Wienert FE (Hrsg) Jenseits des Rubikons. Der Wille in den Humanwissenschaften. Springer, Berlin, S 121–142

Herrmann C (2009) Determiniert – und trotzdem frei. Gehirn & Geist 11:52–57

Janich P (2006) Der Streit der Welt- und Menschenbilder in der Hirnforschung. In: Struma D (Hrsg) Philosophie und Neurowissenschaften. Suhrkamp, Frankfurt

Lesch H (2016) Die Elemente, Naturphilosophie, Relativitätstheorie & Quantenmechanik. uni auditorium. Komplett-Media, Grünwald

Libet B (1985) Unconscious cerebral initiative and the role of conscious will in voluntary action. Behav Brain Sci 8:529–539

Libet Experiment: https://de.wikipedia-org/wiki/Libet-Experiment. Accessed 10 Mar 2019

Mackie JL (1975) The cement of the universe, philosophical books, Bd 16. Clarendon, Oxford

Muller RA (2016) Now – the physics of time. Norton, New York

Nida-Rümelin J, Rath B, Schulenburg J (2012) Risikoethik. De Gruyter, Berlin

Opera "Der Rosenkavalier". https://emilybezar.bandcamp.com/track/die-zeit-die-ist-ein-sonderbar-ding-time-is-weird-richard-strauss. Accessed: 6. März 2019

Penzlin H (2014) Das Phänomen Leben, Grundfragen der Theoretischen Biologie. Springer Spektrum, Heidelberg

Plank J, Bülichen D, Tiemeyer C (2012) Der Unfall auf der Ölbohrung von BP – Welche Rolle spielte die Zementierung? TUM, Lehrstuhl für Bauchemie, Garching

Rasmussen J (1986) Information processing and human machine interaction. North Holland, New York

Romero F (2009) Learning from Flight 1549: How to Land on Water. TIME, Saturday, 17. Jan. 2009

Rovelli C (2016) Die Wirklichkeit, die nicht so ist, wie es scheint. Rowohlt, Reinbek

Schrödinger E (1989) Was ist Leben? Die lebende Zelle mit den Augen des Physikers betrachtet. Piper, München, S 130

Searle JR (2006) Geist. Suhrkamp, Frankfurt

Sheldracke R (2015) Der Wissenschaftswahn. Warum der Materialismus ausgedient hat. Droemer, München

Smolin L (2015) Im Universum der Zeit. Pantheon, München

Struma D (2006) Ausdruck von Freiheit. Über Neurowissenschaften und die menschliche Lebensform. In: Struma D (Hrsg) Philosophie und Neurowissenschaften, Suhrkamp, Frankfurt

Time on the Magic Mountain. https://laemmchen.blog/2016/01/17/die-zeit-auf-dem-zauberberg/. Accessed: 6. März 2019

Vaas R (2016) Die vertrackte Gegenwart. bild der wissenschaft 12–2016, Leinfelden-Echterdingen, S. 30–35

Whitehead AN (1979) Prozess und Realität aus dem Englischen von Hans Günter Holm. Suhrkamp, Frankfurt a. M

Witzleben F (1997) Bewusstsein und Handlung. Ficht-Studien-Supplementa, Bd 9. Rodopi, Amsterdam

Zeh HD (2001) The thysical basis of the direction of time, 4. Aufl. Springer, Berlin

Evaluation and Outlook

4

4.1　Natural Sciences and Humanities in Conflict

We have recalled the ballad "The Sorcerer's Apprentice", described the Chernobyl and Fukushima Daiichi reactor disasters and the explosion of the Deepwater Horizon drilling platform. And concluded that unintended consequences from the use of technical systems are an inevitable concomitant. We have not been able to come to terms with this unsatisfactory explanation and therefore want to ask ourselves the obvious question of directorial control. We were also unable to approach this question directly because of its complexity. We have therefore adopted a deductive approach to solving it, in that, starting from the basic model of the world, we have looked at the cause-effect structure and the spatiotemporal structure. We have mirrored both structures on intentional structures and brought them together. Using the natural science and humanities approach described above, we analyzed the three individual events presented. We remember that we eliminated the ballad "The Sorcerer's Apprentice" because the cause-effect structure could not be applied to it.

This led us to frame the relationship between the natural science and humanities approaches in two questions: "If the natural science side asks: how can there be reasonable causes in a world of causes, while the humanities side asks: how can there be causes in a world of reasonable causes?" (Janich 2006). For the natural science approach, the cause-effect structure and the spatiotemporal structure were used as explanatory attempts to answer these two, admittedly pointed, questions.

We have retained the restriction to the three remaining individual events. The three individual events, however, were additionally examined in a humanities approach from the cognitive perspective according to the model of volitional action, with the formation of an intention to act, in order to come closer to answering the question of reliability and thus to the question of the direction of human action, which is of interest to us.

It is the purpose of this book to relate freedom of action at the individual, group, and societal levels, with their respective constraints at these levels, to the assignment of causality by human action. For causality, we have introduced the physical unit of entropy. Why entropy in particular? Entropy consists of two parts, an irreversible part that we cannot influence and a reversible part that we can shape through our actions. The Second Main Theorem of Thermodynamics thus gives us the possibility of shaping it.

The fact that we have two successful theories, general relativity and quantum mechanics, is the starting point for our considerations. Quantum mechanics and general relativity form the two great advances in physics of the twentieth century. Quantum mechanics has identified the elementary building blocks of our world and how technology can be constructed, realized and used with these elementary building blocks. This process is illustrated with the thermodynamic and psychological arrows of time. The psychological arrow of time (see Fig. 3.2) describes our subjective distinction between past and future events. We can remember the past, but not the future. The thermodynamic time arrow (see Fig. 3.2) is based on the Second Main Theorem of Thermodynamics: The future is the time direction in which entropy increases.

Quantum mechanics describes the world in the Very Smallest – the Entirety Smallest.

General relativity allows an incredible possibility to calculate the entire universe – the Very Largest. The cosmological arrow of time illustrates this phenomenon. The universe began with the Big Bang and may have been expanding ever since. Whether it will expand for all eternity is not certain. According to currently prevailing calculations and theories, it looks like it will. Thus, past time can be gauged by the size of the universe: The future is the direction of the larger universe.

Prigogine and Stengers (1993), pointing to Neumann, writes that the latter found that there was only one entity that could stand outside quantum theory: human consciousness. The past is connected to the present, as in conventional physics, by causality, but the present is connected to the past by consciousness, see Chap. 2 (Sheldracke 2015). Research is also concerned with the connection between consciousness and quantum mechanics. Recent research, especially by neurophysiologist Stephanie Tuss and Bern (2013) on the aforementioned Libet experiment, confirms that apparently, until the experimenter observes or the subject decides, the entire system, MRI (magnetic resonance spectroscopy) – scanner and computer – in a quantum mechanical superposition state explains something that was previously thought impossible. Schrödinger's famous cat would thus become a virtual reality. The ill-fated cat appears in every book about the conceptual difficulties of quantum mechanics. According to this, it would be our consciousness that would be responsible for the collapse of the wave function and the death of Schrödinger's cat through our observations.

A similar connection between relativity and consciousness is confirmed by Goff (2017). He refers to Arthur Eddington, who was the first to prove Einstein's General Theory of Relativity. Eddington's assumption: matter can possess consciousness, the proof of which is the human brain, says Goff.

The results of Tuss and Bern (2013) and Goff (2017) provide evidence that quantum theory and relativity are related to consciousness. Although we are far from understanding all the connections of quantum mechanics and relativity with consciousness, certainly the brain is matter in the highest form of organization.

That the Very Smallest finds the right fit in the Very Largest, man has to see to that through his conscious action. Or in other words: human action must apply the laws of nature in knowledge of the laws of thought.

This is our understanding of the question of direction. The fact that, despite the fit of the Very Smallest in the Very Largest, the catastrophic effects described could occur, is represented by the "tunnel effect" described by Lesch. The likelihood of it occurring we have reasoned with the increase of entropy, information decay, or diffusion of responsibility.

The tunnel effect finds its counterpart in the "Swiss cheese model", which we discussed in Chap. 1, Fig. 1.1.

Consequentialism involves the probability of a consequence of human action in decision making. Consequentialism is concerned with consequences of action intended to bring about a particular effect through causal intervention. Recall that in consequentialism the concept of causality is not taken from the natural sciences and applied to human action, but is reversed, that is, causality is assigned to human action. Consequentialism also makes use of consciousness by recognizing three types of intentionality. Motivating intentionality extends over the entire cause-effect structure and is fulfilled when the causal effect occurs. Antecedent intentionality (decision) is fulfilled with each action step of the cause-effect structure or is not fulfilled because of probabilistic factors. The accompanying intentionality confirms or corrects the results of the preceding decisions.

In other words, it is about purposes, goals, reasons, desires, intentions, etc., which must not be confused with causes. Intentions are intended to bring about a certain effect through causal intervention. And every effect is preceded by a cause. Thus, consequentialism is a suitable approach to resolve the conflict between the natural sciences and the humanities.

We have not limited our scientific understanding to the cause-effect structure. We have extended it to include the spatiotemporal structure and have thus come to believe that the three arrows of time are fused by our consciousness. Consciousness helps us to explain intentionality and thus to understand how conscious action (intention) can lead to gaps in knowledge, which in the present case led to the catastrophic events. These gaps were evoked by repression of the laws of nature. We concluded that if one wants to achieve goals at a given time, one has to make a choice between different constraints.

Or more generally stated: Man is subject to the processes of nature in his actions. He must obey them. In the case of conflict, as in the three catastrophic events, nature has always won and man has been the loser. Or formulated with the inclusion of time: The future is a combination of causal influences of the past with unpredictable elements – unpredictable not only because it is practically impossible to obtain the data necessary for a precise prediction, but because no data causally connected with our experience exist.

This understanding is so fundamental for us that we want to describe once again in compressed form our scientific understanding of the cause-effect structure and the spatiotemporal continuum for the shaping of intentions of our actions.

4.2   Explanations Through the Cause-Effect Structure

With mathematics and physics we have created a grandiose model that puts certain aspects of the world into a logical, mathematical context and gives us a lot of power over nature, see Chap. 3 (Bräuer 2005). Man-made mathematics ingeniously captures very specific aspects of nature, but not nature itself. Objective causal relations in space and time allow a mathematical treatment that leads to the basic laws of physics. In space and time the effects of physical forces express themselves. It is valid:

Energy (or work, which is physically the same thing) = Force − Displacement
Impulse = Force − Time
Momentum = Mass − Velocity

All three relationships can be summarized in the physical term "effect":

Effect = Force − Displacement − Time
Effect = Impulse − Displacement
Effect = Energy − Time

The term "effect" covers spatiotemporally energy, impulse and force. Thus effect becomes the central concept. Spatiotemporal references allow us to consciously experience an outer and inner world. This addresses the question of direction, which is used to make unconscious world contents conscious.

The interaction of consciousness and physics describes the observed effects. It is the objectivity of our spatial and temporal experience of the world and the causality associated with it.

Hamilton's equations of motion completely determine the future behavior of the system if all the spatial and momentum coordinates are known for a given point in time. This fact expresses the determinism characteristic of classical mechanics. Put another way: It describes the dependence of the action on the spatial and temporal coordinates.

The other equation, the continuity equation, a partial differential equation, is a mathematical expression for a conservation law about certain extensive quantities, which states that the increase of such a quantity within an area of space bounded by a closed surface is equal to the difference of time and discharge through that surface. The continuity equation applies, for example, to probability density in quantum mechanics. Put another way: The continuity equation is based on the statistical relationship between cause and effect.

Both equations can be mathematically combined to a linear equation, the quantum mechanical Schrödinger equation. By resorting to the solutions of the Schrödinger equation, one can read off the values of the probability function at any point, i.e. determine the probability of an effect (experimentally with the quantum interferometer), cf. Chap. 3 (Bräuer 2005).

Just as well known as the Schrödinger equation is the relation:

Energy = mass − (speed of light)^2

It states that mass and energy correspond to each other. The functioning of nuclear reactors is ultimately based on this relationship. Mass is reduced and a corresponding (enormous) amount of energy is released.

The cause-effect structure helps us, by means of a scientific explanation, to identify factors that are causally relevant to the occurrence of events. We have for this purpose a heuristic consisting of: Antecedent phase, Cause, Generation of an internal state, Entry of an external system, Point of no return, Triggering event, Causal principle, Probabilistic influencing factors and Effect, for the three individual events, which revealed the causal relevance of the various factors. We have used partly deterministic and partly probabilistic laws to approach the complexity of the three individual events and explain their timing. We have shown the causally relevant factors that became effective so that the chains of events described in Chap. 1 could become this dramatic reality.

What is decisive is that their interaction brings about a process that is not completely determined by the cause-effect structure, but is nevertheless highly probable and which, at least retrospectively, cannot be reconstructed without including the intentional structure.

Or, to put it another way, the strict cause-and-effect structure means that the probability of an effect must be as great as the probability of the corresponding cause.

The connection between cause and effect, causality, turns out to be ambiguous. Our material reality offers possibilities, one of which we manifest as fact. The basic building blocks of matter, that is, the atoms or electrons or nuclear particles prove to be "non-objective". Their effects are contextual. Sometimes they act like waves and sometimes like particles, depending on how they are observed, cf. Chap. 3 (Bräuer 2005).

The concept of entropy makes it possible to show a way by which the occurrence of the individual events discussed can be opened up.

Entropy, which leads to the increase of disorder, is always bound to the factor time. If we observe a current state of low entropy, i.e. high order, a time later, without energy having flowed into the system in the meantime, the entropy will be higher at the later time. Thus, entropy can also be seen as a connecting element, as causality, for the cause-effect structure.

We have shown through our explanations of the cause-effect structure and the intentional structure that the process by which the physical external world passes into the world of everyday life familiar to consciousness lies outside the realm of physical laws. And thus provided us with an understanding of the three individual events that enables us to answer the question posed at the outset about direction.

To summarize, we quote (Janich 2006):

> Man does not merely show causally explicable behaviour (in the sense of the English behaviour, as it is present, for example, in the reflexes, …; in contrast to "behaviour" in the sense of ways of acting, to which the English conduct corresponds, as, for example, in the chains of actions that a person completes by car on the daily way from home to work). The human being shows that he acts and does so purposefully, and, among other things, masters a semantically substantial and eloquent language – …

The time factor does not appear in the cause-effect structure. It comes into play with the spatiotemporal structure, which we now turn to.

4.3 Explanations Through the Spatiotemporal Structure

In a narrower mathematical sense, space is an area extending in three dimensions (length, width, height) without fixed boundaries – a visual space. This visual space is described by the three-dimensional Euclidean space R^3. Space in physics is a fundamental concept for grasping the mutual arrangement of bodies. The succession and duration of motion sequences and physical processes in space is expressed in time as an ordering parameter, see Chap. 3.

We have pointed out several times that the properties of time can be described but not explained. Therefore, we have described the spatiotemporal structure with the theory of relativity, a summary of the three space dimensions with time as the fourth coordinate to a four-dimensional space (spatiotemporal continuum). The concept of spatiotemporal is an expression of the close connection between space and time in the two theories of relativity, thus enabling their unified description, cf. Chap. 3. Special relativity is based on a quasi-pseudo-Euclidean metric (related to the metre as a unit of measurement), which describes the extension of Euclidean space of classical physics by the temporal dimension to Minkowski space M^4 (plane space-time). In general relativity, the metric is in principle non-Euclidean, and the metric tensor (term used in vector calculus, vectors are physical quantities that include, in addition to magnitude, the specification of direction and sense of direction) depends on the coordinates. General relativity, as a geometric theory of gravity, provides a complete relativistic description of the gravitational field. This leads to a curved spatiotemporal and also later-time (spatiotemporal curvature), see Chap. 3. The general theory of relativity is confirmed in its predictions by a whole series of experimental proofs.

Kant thought that the temporal unity of our ego is an original achievement of our consciousness, a capacity to create unity in the multiplicity of sense experiences. Today's neurophysiology and cognitive psychology prove him right; they show that our perception of time is based on discontinuous processes, so that our experience of "time" in the form of a continuously passing flow of time must be a construct of our consciousness – whatever this consciousness is, which we stubbornly fail to explain, cf. Chap. 2 (Falkenburg 2012). We quote Bräuer in Chap. 3:

> It is like the blind spot of the eye: our perception of the world is flawed, but we do not experience the flaws. Consciousness adjusts with a delay and does all it can to hide the fact – from itself. It deceives. It deceives itself. This is very expedient – given enough time. (Brewer 2005)

Kant's approach helps to reconcile the aforementioned contradiction, but does not fully clarify it. Therefore, let us return again to the mentioned contradiction with the distinction between reasons and causes. Causes are objective; they belong to physical phenomena. Reasons, on the other hand, are subjective; they belong to mental phenomena. As emphasized, one must not confuse physical causes with mental causes.

Reasons can be seen as a conscious form of experience of brain processes.

Causes have nothing to do with the form of experience; they can only be inferred deterministically through the cause-effect structure, cf. Chap. 2 (Penzlin 2014).

Man perceives reality as through a mirror. Or stated differently: In man, the universe views itself as through a mirror. Or reciprocally said: The universe can be understood as a continuous extension of the human body.

People can only influence their environment to the extent that they receive information from their environment. Without information, all creative action would be impossible. As spatiotemporal action progresses, it is consciousness that informs. The universe is unfolded in all its temporal dimension. Time is something given, it does not trickle away. But as a consequence of the adaptation of consciousness to the development of our universe, consciousness can only accompany this process of acquiring information in the direction of increasing entropy and thus the direction of time.

The acquisition of information, the shaping of organization, the compensation of wear and tear on machines, the use of tools that relieve man, etc., require a concentrated use of energy with the purpose of "buying, that is, gaining" time. Every creative act makes it possible to compensate for the passing of time; every newly developed machine creates time, potential time, and potential time is information. With the creation of information, on the other hand, entropy increases, errors thus become more frequent.

Our epoch, so greedy to "buy" time, is characterized by this time-gaining action as much as by the conquest of our space.

One can also consider the events in the spatiotemporal relational framework as an interplay of natural events and actions.

Natural events are determined, they are strictly subject to the cause-effect structure. Actions are rather underdetermined. Under the restrictive conditions of the spatiotemporal relational framework, man can react with his behaviour in a correspondingly "resilient" manner. However, this also implies that man contributes to his own lack of freedom and thus limits his freedom of will itself.

Again, we conclude that it is reasons, attitudes such as desires, fears, opinions, hopes, intentions, etc. that are supposed to help us cope in the causal world, which unfortunately was not the case with the catastrophic events.

4.4 Explanations Through the Intentional Structure

We have shown the conflict between the natural sciences and the humanities from the natural science side through the explanations of the cause-effect structure and the spatiotemporal structure. In this section we will look at this conflict from the point of view of the humanities, using intentional structures.

Teleological (purposive) explanations assume intentions, and these are irreducible (non-derivable) from the standpoint of scientific explanation.

Scientific explanations, whether they are deductive-nomological (deductive-nomological is a formal structure of scientific explanation of a causal relationship using natural language), probabilistic, or based on causally relevant mechanisms, leave the making of our plans, intentions, and actions in the dark. Explanations are either scientific or focused on the goal of action, but not both; one excludes the other, as we have seen by contrasting the cause-effect structure and the intentional structure for the three individual events. This underlines, once again, the importance of the introduced distinction between reasons and causes. Again, reasons are not causes; a strict distinction must be made between the two.

Whether a person has acted for the reasons, purposes – not causes – that he or she assumes, cannot be decided with certainty by the person himself or herself, let alone by another person.

As a physical system, the brain is also subject to probabilistic laws in addition to physical laws.

Probabilistic laws do not allow predictions for individual events, but only for statistical aggregates. All statements about the individual case, at which a probabilistic explanation aims, are afflicted with the problem of induction. They lead to inductive conclusions that are not compelling as to the occurrence of an individual event within a certain period of time. A probabilistic law precisely does not predict if and when a particular event will occur. Also the explanation of a past single event remains incomplete; because why an event occurred at a certain point in time and not earlier or later or not at all remains unexplained (see "tunnel effect").

We cannot treat laws of thought as laws of nature because they are laws to be followed at all costs so that such events as the three catastrophic events described do not occur but, as we have also seen, were not necessarily followed. Following is a recommendation, but not a must. On the other hand, the natural scientist must anticipate the laws of thought before he applies the laws of nature. A physical machine is a construction following the laws of nature, but the constructor cannot entirely anticipate the background of the intentional action of the operator.

Bennett and Hacker (2006) state:

> Theories and hypotheses in the scientific sense do not occur in philosophy. For in the sciences theories serve to explain phenomena, and hypotheses are used to explain them. It must be possible to test scientific theories in experience. They may be true (or false), but they may just as well be approximations to the truth. Philosophy, on the other hand, clarifies what is meaningful and what is not. Determinations of meaning precede experience and are presupposed by true judgments as much as by false ones. In philosophy there can be no theories from which one can derive hypotheses about events or with the help of which one can explain why things happen as they do in fact.

Janich (2006) addresses the difference between the natural sciences and the humanities by stating:

> The natural scientific and the spiritual scientific view of man are mutually exclusive: Man ... in the perspective of the natural sciences is ultimately also only matter functioning causally according to natural law; in the perspective of the humanities, the cognizing human being ... remains dependent on cultural achievements ... in his knowledge of mind and brain.

We will explore this finding in the context of addressing organizations, their corporate culture or specific safety culture.

We still stay with Janich because we share his remarks on the difference between reasons and causes: "Or, in short, in a world of causes there are no (independent of them) reasons, and in a world of reasons there are no (independent of them in the sense of unrecognized) causes" (Janich 2006).

These two limitations show us a way out to resolve the conflict. In place of the conflict, we place a meaningful kind of complementarity and cooperation between aspects of the natural sciences and the humanities.

In conclusion: We cannot explain intentional structures of actions by the fact that someone wants something.

4.5   Resolution of the Conflict Between the Natural Sciences and the Humanities

This summarizing statement leads us to the mentioned double character of actions. The dual character is formed on the one hand by the cause-effect structure and the spatiotemporal behaviour and on the other hand by the consequentialism coined by Nida-Rümelin. For the shaping of consequentialism stands the volitional action of man through the pursuit of goals (consequences), the implementation of plans and intentions into action. Free will is limited by external circumstances and other decision-makers, but also by the agent himself. We have seen that for the explanation of the three catastrophic individual events the cause-effect structure and the spatiotemporal behaviour are not sufficient. In the explanations we also had to include the intentions and thus the reasons or some other driving force, as is not done with the cause-effect structure and the spatiotemporal behavior. The freedom of action, the driving force of the agent, and the intentional or ordered decision to cross the "Rubicon" are completely different from the elements of the two structures of natural science. Determinism excludes freedom of will. In contrast, free will is a reality in rule- and knowledge-based behavior. We recall Searle (2006) in Chap. 2 saying that consciousness organizes the brain so that conscious decision-making can occur even in the absence of causally sufficient conditions. This was not the case with the archer. He made the decision to release the bow tension and thus passed the Point of no Return. Thus, the causal condition for the triggering event, the consummation of the decision, was satisfied. In the Chernobyl reactor disaster, on the other hand, the causally sufficient conditions for the decision to continue the experiment after a half-day interruption were lacking because they were not met by the changes in the neutron physical state after the Point of no Return was exceeded during the interruption of the experiment. At Fukushima Daichii, the plant was operated with an inadequate and unprotected emergency power system, causing the Point of no Return to be exceeded. The initiating event, the tsunami wave, indicated that the causal conditions for the operation of the power plant were not in place. On the Deepwater Horizon rig, the Point of no Return was formed by the wrong drilling technique. The premature replacement of the drilling fluid overlying the cement with seawater in the wellbore showed that the causal conditions for the drilling crew's decision to make the replacement were lacking.

One could summarize it like this: All three individual events clearly show that the respective intentions led to a failure to recognize that the Point of no Return had been exceeded, or that the triggering event was brought about deliberately, or could be brought about by a natural event. The latter happened at Fukushima Daiichi.

In Nida-Rümelin's Consequentialism in Chap. 2 (Nida-Rümelin et al. 2012) we find an approach to overcome the contradiction. He formulates that in which way one describes a concrete behaviour at a point in time in a certain place as an action depends on both the causality of the action and the spatiotemporal behaviour and on the intentions of the agent. We regard this observation as crucial, because according to it there is not only one goal-directed intention for an action, but in his model it is formed from different intentions that act causally and spatiotemporally as well as mentally. Nida-Rümelin thus resolves the contradiction by considering action-constitutive intentionality as complex. We have dealt with this complexity in Sect. 2.9, drawing on the Rubicon model of action (Fig. 2.3, Chap. 2). According to this model, intention includes the causal and spatiotemporal character of action and the mental (more specifically, intentional) character of action. We have noted that the reason for action (motivating intention) is to the left of the Rubicon in the Rubicon model of action, and the preceding decision is formed to the left of the Rubicon and leads to the crossing of the Rubicon with the central decision. The accompanying intentions (behavioral control) extend to the three further steps after crossing the Rubicon in the Rubicon model. To clarify, actions consist of the cause-effect struc-ture and the motivating intention (both terms in the singular). In addition, actions are shaped by the spatiotemporal behavior and the accompanying intentions (behavioral control) (both terms in the plural). Here, we understand the accompanying intentions as behavioral controls. The behavioral controls refer to the whole action and consist of several self-contained control loops according to the Rubicon model of action. We have explained the differentiation of action or behavior and intention with the three individual events and found that the catastrophic end state in the individual events can be attributed to a deficiency in the accompanying behavioral control and that the motivating intention was not fulfilled by the catastrophic effect at the end of the event. The preceding (action-guiding) intentions are fulfilled by each action step and have an influence on the spatiotemporal behavior during the respective action step. These connections are to be represented graphically here, see Fig. 4.1.

Upper level: Cause and effect structure

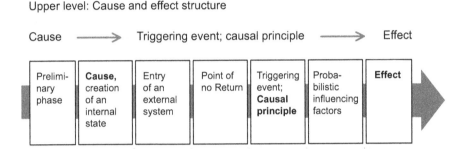

Cause ⟶ Triggering event; causal principle ⟶ Effect

| Prelimi-nary phase | **Cause,** creation of an internal state | Entry of an external system | Point of no Return | Triggering event; **Causal principle** | Proba-bilistic influencing factors | **Effect** |

Motivating Intentionality (reason for action) is fulfilled with the onset of the causal effect or fulfilled or not fulfilled due to probabilistic influencing factors.

Previous intentions (decisions) are fulfilled with every action step.
There is no probabilistic blur here.

Accompanying intentions (behavioral control) confirm or correct the results of the preceding intentions (decisions) of each action step. The behavior control is carried out with each action step and influences the spatiotemporal behavior of the action step.

Lower level: Consequential reasons for action

Fig. 4.1 Interaction of cause-effect structure, spatiotemporal behaviour and mental intentions

As shown, motivational intentionality extends over the entire cause-effect structure. The preceding intentions find their confirmation with the completion of the respective action step and end with the triggering event. The accompanying behavioral control is a control loop for each action step and ends with the Point of no Return. This is also true for spatiotemporal behavior. At the triggering event, intervention is no longer possible, the event takes its course.

We still remain with the interaction of nature and spirit. In antiquity, the formula "neither of nature nor against nature" applied. This formula states that man is not only subject to the causes of nature, but that his reasons for action can also follow principles set by himself under given conditions. Within this framework, man can shape his life. Struma (2006) describes it this way:

> The special significance of the human form of life is rooted in the superimposition of artificial orders on the natural order. Persons are not only exposed to external constraints and influences, but can indirectly initiate their own behaviours through the development and establishment of artificial orders. ... The ability to establish artificial orders in the world nevertheless distinguishes the human form of life in an eminent way from other animal forms of life.

Persons are equally subject to causes and susceptible to reasons, purposes, goals, etc.

The concept of free will presented here is not limitless, but pursues the goals set with consequentialism. Freedom of will does not assume an opposition of freedom and determination. Human freedom manifests itself as inner determination by reasons (Struma 2006).

The Libet experiments we have described in no way contradict the previous understanding of free will. It is true that Libet interprets his experiments as evidence that no free act of will initiates the actions (cf. the causal chain bow shot in Chap. 2). He does concede, however, that persons at least have the possibility of stopping the movements set in motion by the brain. This veto function, which he also outlines, obviously fulfils control functions in action sequences (Struma 2006).

We conclude this section with a quote from Einstein: "The intuitive mind is a sacred gift and the rational mind its faithful servant. We have created a society that reveres the servant and has forgotten the gift."

To see nature and society "thermodynamically" is to understand the competition between structure formation and decay. The formation of structure is reflected in the striving for self-realization and in the exhaustion of free will. The two are in conflict. Self-realization goes back to individual dispositions. Freedom of will is unfolded or limited by the "webbing" society grants to self-realization. This creates the potential for conflict. Society forms the overarching system for free will, which must defend itself against entropic decay. This happens through the cultural framework for living together in society. This framework, on the other hand, leads to the restriction of free will, since individual and social interests often contradict each other. The art now consists in bringing these conflicting interests into harmony with each other through decision-making.

In Sect. 4.6 we will show what constraints man is subject to in dealing with this gift. Man has the possibility of changing his behaviour under the restrictive conditions imposed on him by his environment (nature), his own and other people's desires, goals, etc. This also includes, however, that people contribute to their own lack of freedom and against their better insights and reasons. But this also includes people contributing to their own lack of freedom and acting against their better insights and reasons, as in the case of the three catastrophic individual events. We want to explain these limitations on free will through the mechanisms of choices.

4.6   Decide

We would like to start this section with an event from the everyday working life of production in the automotive supply industry.

This event relates to the manufacture of airbags. The airbag (airbag) is a component of passive safety in the automobile.

The airbag fulfils its task through the following functional chain:

1. At time "zero" the crash happens.
2. 25 ms (milliseconds) later, the electronic sensor activates the driver module's firing pill (we are limiting ourselves to the driver, i.e. other airbags in the vehicle function according to the same principle).
3. After 30 ms the cover of the driver's module is torn open and the airbag is inflated.
4. After approx. 55 ms, the driver airbag is fully inflated and the driver dives in.
5. After 85 ms the driver is maximally immersed in the airbag and moves away from the steering wheel again.
6. After 150 ms, the entire accident is complete, the occupants are in the starting position and the airbag(s) are largely deflated.

There are various airbag units. Our example below relates to airbag units with hybrid gas generators consisting of a pressure vessel containing a compressed gas and a pyrotechnic assembly used to trigger and heat the outflowing gas. The following event relates to the manufacture of such airbag units.

During the assembly of the airbags, it was noticed by slightly cooling the hand that gas was leaking from the pressure vessel. The fitter informed a person he trusted of this, who in turn informed the decision-maker ultimately responsible of these observations. This decision-maker decided to continue the assembly of the airbags without clarifying the observed cooling of the palm and to install the defective pressure vessels, although in the meantime the feeling of the fitter was confirmed by bubble tests in the water (rising bubbles) that the pressure vessel was slowly but steadily emptying.

We do not wish to comment on this observation here, let alone extrapolate, but rather to use it to demonstrate the mechanisms of decision-making.

Before turning to the mechanisms of decision-making, we would like to point out that the observation outlined shows how difficult it is to demonstrate formal, legally secured participation in the decision-making process. Similarly, the communication process leading to that decision is hard to demonstrate because it occurred apart from the written organizational channels. Increasingly specialised knowledge for the work process makes the decision-making scope of employees and decision-makers more flexible. This makes it necessary to have a constant dialogue between managers, their employees and the management. The dialogue can be patriarchal (key words: "par ordre du mufti" and "ukase"), i.e. one-dimensional, or cooperative (key word Latin "pater familiae", literally "father of the family"), i.e. multi-dimensional. We have already explained the expression "par ordre du mufti", but we want to add that such decisions and instructions based on authority always have a demotivating effect on those concerned. Sometimes, however, they are unavoidable. It would then be necessary to examine in detail the extent to which the leadership structure in place in each case has served to prevent dissent from those affected. We need to explain the term "pater familiae". Although the "pater familiae", as a leader, has the final decision-making power and consequently also has to bear the undivided responsibility, he is bound by law, custom, corporate culture, habit and ethics, morality. The leader can only live up to this responsibility if he follows the advice of the "familia", cultivates and promotes consultation and cooperation with the "familia", i.e. practices a cooperative style of leadership.

Leadership style, regardless of its form, is ultimately about influence or the possibility (power) to do so. The root of the word "power" does not come from "to make", but from "to like". Power is thus the possibility or potency (the ability!) to set something in motion. Thus the concept of power touches upon that of leadership, in which Dahm's (1963) "movement effect" is a crucial defining feature. We have described the "movement effect" with the three arrows of time, the thermodynamic, the psychological and the cosmological.

The concept of power draws attention to circumstances that are to be localized in the organization. Power serves to absorb uncertainty, which is to be avoided in the company. In this context, the power profile of an organization does not coincide with the formal superordination and subordination in a company.

As an everyday language term, power cannot be narrowed down to a single one of its many uses. We use the term power in the sense of causality (cause-effect structure), control (spatiotemporal behaviour), motivation (intentional structure) and causation (consequential reasons for action). Power serves to control information in organizations. Through this process of controlling information, the organizational system differs from the previous state. We have described this difference of system state at different points of time as entropy increase. Or as stated in detail: Information processing is generation of negentropy.

Or to put it differently: "Decisions can only be communicated if rejected possibilities are also communicated, because otherwise it would not become comprehensible that it is a decision at all (Luhman 2006)".

We stay with Luhmann and capture the phenomenon of time with another quote from him:

> Decisions mark a difference between past and future brought about by themselves. They thus mark the irreversibility of time. Remarkably, this takes the form of events that are themselves bound to points in time, that is, they are neither reversible nor irreversible. (Luhman 2006)

Every decision presupposes the "universal time" shaped by the thermodynamic and cosmological arrow of time, which continuously transfers the distinction between past and future into another, a new present (Luhman 2006). This means, of course, to take the present seriously as the place of real decisions, thus also of the struggle for this decision with regard to a future that has obviously become risky for all.

Causality is also associated with time, because causes must occur earlier before effects can occur. Luhmann states: "Elementary causality in the sense of an effecting of effects by causes is therefore always a time-consuming event that bridges time differences (Luhman 2006)".

"Causality thus seems to be nothing other than time schematized in a certain way, just as it is in a different way for space," Luhmann continues (2006).

We have shown that decisions led to the disastrous results in the three individual events.

Again in a nutshell: For the RBMK reactor at Chernobyl, the decision was made to go through with the planned test by "hook or crook". For the Fukushima Daiichi plant, the decision was made to build the plant in a coastal region vulnerable to tsunami and to operate it without adequate safeguards. On the Deepwater Horizon rig, the decision was made to use the wrong drilling technique to absorb deadline and cost pressures.

Decisions are often accepted because one does not want to hurt the decision maker. The decision-maker is thus also dependent on the decision-recipient. Never, as shown by our observation above, would an exploration of the real events in a goal-directed way be allowed in the system, because such an exploration would bring the system to a standstill, moreover it fails because one cannot determine how decisions "really" come about. Moreover, there is the fact that every decision presupposes the "now", which stands between the past and the future. In the present of the decision, the past provides disposable resources that can be used differently. The unknown future is brought in as an intention. Such open decision-making premises are accepted and "channeled" through organizational culture. Classical organizational theory, on the other hand, assumes that the organization produces the decisions. It invokes the official job descriptions, according to which the organization is concerned with decision-making at the rule- and knowledge-based level. But in fact it is about transforming an unknown situation into the fixed world of the organization. And this can only be done by people within the framework of the principles of the organization's culture. As our observation above and the three individual events show, decisions can only be made currently, always in the present, and are therefore affected by the temporal "uncertain" process. Decisions themselves, however, remain a mystery – but a familiar and well known one that can be experienced every day (see airbag). One therefore sees decisions as a "qualitas occulta, hidden quality" or human contribution – all three terms stand for properties, hidden qualities of events (Luhman 2006).

Could one also speak of randomness of the decision?

Or, according to Daniel Kahneman's decision theory, that decision-makers in general and managers in particular cannot follow the rules of the models of rational decision-making; their daily work simply does not allow them the time to do so. Management overplays this time pressure by a "special style" of proclaiming decisions to themselves and to the executors. This brings us to the core of the problem of decisions. Decision-making is about the problem of assessing risk, which relies on the associated observations, like our observations of the manufacture of the airbag, i.e., second-order observations. The decisions that ultimately determine the further course of action are attributed to collective memory, even if they were made by individuals. Decisions are tied to power, and this requires a differentiated functional system, a hierarchical order.

We quote Luhmann (2006) again: "Decisions are only possible because the future is indeterminate, that is, unknown. And this is precisely what constitutes what is usually called the future." We note that the unknownness of the future is an indispensable condition for the decision-making process. It is also what makes answering the why question so difficult. The organization – meaning the company – ultimately relies on individual memory, and this often leads to divergent answers because the people involved in the decision communicate with each other against a background of differing perceptions. When reconstructing events, it is therefore better not to ask the why question, but to interpret the event on the basis of the cause-effect structure, the spatiotemporal, the intentional and the consequential behaviour.

A system memory does not exist, just as a corporate knowledge does not exist; both draw on the personal memory and knowledge of their members. This is confirmed by the memory and knowledge functions necessary for risk-informed decisions that must be made by the decision maker(s). Such decisions must absorb uncertainties that can only be weighed by the decision maker(s). From this we conclude that there are no uniquely correct decisions and that decisions about decision premises (see antecedent phase in the cause-effect structure and motivating intention of the intentional structure) must adjust to memory and to any correction of decisions. Decision, knowledge and memory thus mutually condition each other (Luhmann 2006). Memory depends on the support of consciousness. The latter, however, stubbornly refuses any scientifically closed explanation, as we have noted. We would like to recall that conscious action is required at the rule- and knowledge-based cognitive-behavioural level. Decisions at these two levels aim at the difference between past and future. Without decisions, there would be no predictable action at these two levels, i.e., no planning of behavior even in the case of counteracting wrong decisions. Decisions also do not change the unknown of the future. This problem can also not be solved by obtaining more information, because this is always sought in the past and differentiates the unknownness of the future, but does not eliminate it (Luhmann 2006). It is precisely this uncertainty that is expressed by the term risk. The concept of risk (in contrast to the concept of danger) denotes a form of observation of decisions (Luhmann 2006). "Risk consciousness is thus a functionally equivalent attitude toward the future by which decision-making makes itself possible in the first place; ... (Luhmann 2006)."

We would also like to present this explanation of Luhmann in a different way: If this uncertainty were not present in decisions, devastating consequences would occur if the future were known. This would make any kind of decisions superfluous, because what happens is what is determined. Decisions are, of course, subject to criticism as we have raised them in the analysis of the three individual events. This brings us to the subject of responsibility. The decision-maker bears responsibility for his judgment, his criteria of judgment, and his assessment of the situation at the time of his decision. Again, this responsibility must always be tailored to individuals. Because of the time-bound nature of the decision, they can only make the decision to minimise the risk once it has been recognised that the Point of no Return has been exceeded and could then take appropriate countermeasures. But all this, with the exception of archery, did not happen in the causal chains described.

The archer made the decision to release the bow tension, thus he brought about the Point of no Return, only with the execution of the decision the archer brought about the triggering event, the irreversibility for the hit "into the black". In Chernobyl the last remaining safety system was switched off by the operational personnel after the instruction of the load dispatcher (distributor of electric power within a certain territory), the Point of no Return was set. The decision of the load dispatcher to continue the test after an interruption of about half a day created the triggering event, the catastrophe was inevitable.

The Fukushima Daiichi plant was operated without adequate protection of the independent emergency power supply, following a management decision. The tsunami wave, the triggering event, exposed this design error.

On the Deepwater Horizon rig, the drilling crew decided to add too much retardant to the drilling fluid, passing the Point of no Return. Another bad decision by the drilling crew, the premature replacement of the drilling fluid with seawater, led to the explosive gas cloud ignited by an accidental spark, the triggering event.

Summary

Erroneous decisions led to the Point of no Return being exceeded in the three individual events.

Or to put it another way: The decision-makers lacked risk awareness at the time of their decision. In all three cases, this was due to the deadlines that had been set. The tighter the time planning and the tighter the deadlines, the more susceptible the system is to disruptions. Time pressure influences the cognitive quality of information processing and also leads to a tendency to "outsmart" safety regulations if their observance is perceived as disproportionate in the specific situation.

Under time pressure, information that is easy to obtain is preferred over information that is difficult to obtain. Preference is also given to quick decision-making authority instead of thorough analysis of the problem, which of course takes time. This also creates time pressure, which has a discouraging effect, which leads to an uncomfortable suspicion being suppressed or pushed aside.

The destruction of the three plants further illustrates the need for time to gather information in order to make a decision tied to a point in time.

We would like to present the influence of decision in more detail on the basis of the three individual events before we deal with the repeatedly mentioned aspect of corporate culture.

4.6.1 Chernobyl

The cause-effect structure is represented by the neutron-physical behaviour of the graphite-moderated boiling water pressure tube reactor, to which the operator obviously did not pay sufficient attention during the experiment. In the course of the experiment, the neutron population led to a "supercritical" (neutron excess) and thus uncontrollable reactor.

The spatiotemporal relationship frame can be explained by the "exercise of power" ("par ordre du mufti") by the load dispatcher. The fulfilment of his instruction led to the shutdown of the last remaining security system. A repetition of the test would only have been possible after several years, because the test requires a special neutron-physical configuration of the reactor core.

The intentional structure (here in particular the pre-run phase) was characterised by the experimental objective to prove that the rotational energy of the turbine set being phased out was sufficient to supply the main coolant pumps with electrical energy for post-cooling purposes in the event of an unscheduled grid shutdown with rapid reactor shutdown (fulfilment of the five-year plan).

The consequent reasons for action were created by the instructions of the load dispatcher, whose task is to maintain the electrical power supply of Ukraine, even during the execution of the experiment. In short, the control of the test procedure was completely in the hands of the load dispatcher, and thus the operators of the reactor were degraded to "command receivers".

We must again detail the preliminary phase and the consequential reasons for action. On the preliminary phase. The RBMK reactors were chosen because of their economic efficiency, in accordance with the decisions of the Soviet Union Ministers of June 19, 1969 and December 14, 1970. Construction of Chernobyl Unit 4 (the unit destroyed by the disaster) started on April 1, 1979, completed in 1983, grid synchronization on December 22, 1983, start of commercial operation on March 26, 1984. Consequential reasons for action: The commissioning programme also included the experimental validation of the accident concept through the test as carried out on 26 April 1986. However, it was not carried out during commissioning because otherwise the deadline for grid synchronisation (22 December 1983), in accordance with the five-year plan, would not have been met. The decision to carry out the test at a later date resulted in the commissioning of the reactor, whose safety case was not complete. This decision directly led to a "sword of Damocles" that constantly hovered over the plant. To the further wrong decisions and the instructions of the load dispatcher during the trial on April 26, 1986, we must add another aspect. The instructions were by no means received by the operational personnel without "grumbling". Only by the threat of dismissal (pure blackmail) the operative personnel followed the instructions which were incomprehensible to them.

Thus, the question of direction and reliability for the Chernobyl reactor catastrophe can be answered unambiguously: it is the directives based on wrong decisions.

A strategy and practice of state supervision is not discernible.

It should be noted that in 2017, four first generation RBMK reactors, six second generation RBMK reactors and one third generation RBMK reactor were still in operation after a large number of modifications, which affected the reactor shutdown systems in particular. In the long term, it is also planned to replace the pressure tubes in the reactor core.

4.6.2 Fukushima Daiichi

The cause-effect structure is described by the inadequate state of protection of the plant. The plant was not sufficiently protected against plant-internal accidents and against external events. The design concept was neither probabilistically nor deterministically balanced.

The earthquake (tsunami) hit an inadequately protected plant that was built in a region that was already considered to be at risk from tsunamis in the past. The 41 min available to the operator as a spatiotemporal reference frame to be prepared for the failure of the emergency power supply were obviously not used due to the lack of instructions for action. (41 min passed from the registration of the earthquake off the east coast of Honshu to the arrival of the tsunami wave at the power plant site). After that, the room for manoeuvre had shrunk to zero because the plant had practically no electrical power due to flooding.

The intentional structure is represented by the economic operation of a plant. A characteristic element of the practiced intentional structure is the lack of accountability, which leaves the question of responsibility in the dark, an element typical for the Japanese management culture. It led to complacency, which was also supported by unquestioned operating results.

The earthquake hazard was known, as was the endangered location. Expert opinions and counter expert opinions were supposed to "mitigate" the danger. Here, too, "convincing" operating results meant that there was no reason to question the operating result for consistent reasons for action.

The decisions during the construction and operation of the Fukushima Daichii plant, which are mainly in the preliminary phase of the cause-effect structure as well as in the motivating intention of the intentional structure, are due to psychological factors that help people protect their beliefs and actions, and thus their self-actualization. The earthquake and tsunami risks were obviously underestimated. This fact is – in hindsight – undisputed and, at least as far as the tsunami risk is concerned, confirmed by both the IAEA (International Atomic Energy Agency) and the Japanese government, see Chap. 2, (The National Diet of Japan 2012). Instead, the risk calculations were based on historical data (see Fig. 3.6 in Chap. 3). Paradoxically, these historical data were used selectively. For example, there was a huge tsunami wave in the aftermath of the Jogan earthquake in 869 that exceeded that of the Fukushima Daiichi design, see Chap. 1 (Mohrbach 2012). The other decisions during the operation were due to psychological displacement and selective perception. The typical Japanese decision-making culture, which is characterised by the fact that it remains anonymous and is made collectively, i.e. would never have been made by one person alone, also contributed to this.

The strategy and practice of supervision in Japan can be described as a non-transparent conglomerate.

The question of direction and reliability can also be fixed here in the human weaknesses.

4.6.3   Deepwater Horizon

During the development of an oil reservoir with a reservoir pressure of approx. 900 bar, the cause-effect structure was formed with improper sinking of the well and with the connivance of the supervisory authorities.

The spatiotemporal frame of reference was limited to the drilling platform. The ignition of the explosive gas cloud by an accidental ignition spark should have been part of a safety analysis, which was obviously omitted due to the enormous cost and time pressure.

The intentional structure was characterized by a negligent "outsmarting" of safety requirements, which was expected to "buy time" by shortcutting work

Table 4.1 Summary of our analyses of the three individual events

	Cause-effect structure	Spatiotemporal behaviour	Intentional structure	Consequential reasons for action
RBMK reactor Chernobyl	Thermodynamic and neutron-physical behaviour of the RBMK reactor ignored	The planned experiment could only be carried out within the framework of a planned shutdown. A repletion would only have been possible after years	Test to prove that the rotational energy of the expiring turbine set is sufficient to ensure post-colling in the event unscheduled grid-failure (Five-year-plan)	Control of the test sequence by the instructions of the person responsible for the distribution of the load in the electrical supply network (load dispatcher)
Fukushima Daiichi nuclear power plant facility	Insufficient protection of the emergency power supply system against tsunami waves. The emergency power system was also inadequate in terms of its capacity	An earthquake off the east coast of Honshu with a magnitude of 9.0 caused a tsunami wave. It caused flooding because the power plant was built in a vulnerable area	Operation of a nuclear power plant, which did not meet the safety standard considered necessary internationally	The earthquake hazard was known, and attempts were made to "mitigate" it through a variety of expert opinion. Satisfaction with the previous operating results meant that no safety improvements were made
Deepwater Horizon drilling platform	Development of an oil storage site with a pressure of approx. 900 bar with inadequate drilling technology	Apparently, a spark from a by chance passing ship caused the gas cloud to explode and destroy the rig	The delay of 43 days and the accrued costs of approx. 30 million US$ were to be absorbed	Use of a non-functional Blowout preventer and Some bad decisions in replacing the drilling fluid above the cement with seawater

procedures in order to compensate for the costs of about US$30 million accumulated due to delays of 43 days.

Consequential causes of action can be seen as the use of a non-functional blowout preventer and a number of bad decisions in replacing the drilling fluid above the cement in the well with seawater, all with the approval of the prospecting authorities.

The decisions taken during the course of the incident were due to various serious failings. This is also the view of the US Department of Justice, which imposed a fine of US$4.5 billion as a result of the oil spill. The highest fine ever imposed for an environmental offence.

The decisions made by the rig crew were apparently motivated by financial savings.

The role of the supervisory authority is illustrated by the following fact: the supervisory authority refrained from drawing up the prescribed emergency plans for accidents on the grounds that they were improbable to impossible.

Table 4.1 shows the results of the analysis of the three accident events.

We recognize that in causal constellations, decision-making on the basis of consequential reasons for action plays the dominant role. Decisions made on the basis of consequential reasons for action allow causality to take effect. We find Nida-Rümelin's approach about the influence of reasons for action on the cause-effect structure confirmed. We also find confirmed the coupling of causality and time shown in Chap. 2. Causality is the decision that establishes a connection between cause and effect, that is, that uses the medium of causality (Luhmann 2006). With every decision, causality is established without being able to generate the necessary causes and the resulting effects itself (Luhmann 2006).

The freedom and autonomy of actions must take into account an appropriate dosage of existing constraints imposed by the decisions. If these constraints are pushed aside by the decision-makers, for whatever reason, there is a danger of moving from a state in which accepted and controlled safety constraints are effective to a spatiotemporal frame of reference that can no longer be controlled and thus cannot be controlled, as happened in the case of the three individual events, leading to the destruction of the entire system.

In summarising our analyses of the three individual events, we have highlighted the dominant role of the decision and also addressed the steering of decisions by corporate culture. Not only is corporate culture a corrective to decision-making, but there are other aspects that we will present in Sect. 4.7.

Before we focus on the decision aspects, we would like to recall the connection of the Very Smallest in the Very Largest presented by Lesch (2016) in Chap. 2. We have explained this connection in terms of the single-mindedness of the three arrows of time. We subsume decisions, and thus decision aspects, under the concept of the arrows of time because they reveal the typical single-mindedness that characterizes the development of all life. This single-mindedness is also found in normal every-day consciousness, especially when making decisions, because decisions presuppose previous events and anticipate the unknown future. Aspects of decision-making are the precondition for decision-making and the result of the interaction of self-realization and freedom of will, our everyday challenge. This everyday challenge, the conscious decision for an alternative course of action, is as much a part of the very small as the quantum mechanical processes mentioned above.

We therefore conceive of decision aspects as a tool available to humans for shaping the Very Smallest of things.

4.7 The Very Smallest

The elements of the Very Smallest, the decision aspects, are based on the results of our field research, they go back to our participant observation. They are therefore not exhaustive. Field research is understood as a systematic exploration of cultures or specific groups by going into their habitat and directly experiencing the everyday life of this group of people.

We draw on the considerations of Hoensch (2006), which he has underpinned with results from his observations of nuclear power plant personnel. We would like to start by explaining the concept of corporate culture that we have already introduced.

Basically, it can be said that the decision models include three components, namely factual, which is directed to results, evaluative, which is goal-oriented, and methodological. The latter includes algorithms (computational methods), methods that are fully operationally described, and heuristics, methods that are not algorithms.

We do not want to pursue these basic structures further here, but turn to the aspects that we found out during the participant observations.

4.7.1 Corporate Culture

The word corporate culture could superficially be seen as a fashionable topic. Today, the meaning of this word and especially the association with the positively connoted (with a word associated additional idea) concept of "culture" seems to fade somewhat. By the mid-1980s, voices were already being heard declaring this term to be over. Due to the Chernobyl reactor disaster, corporate culture received a "reframing". This organisational evolution goes along with (Luhmann 2006) the loss of (or renunciation of) centralised control, preference for informal contacts, soft divisions and categorisations, loose linkages, non-transparent network formations, greater dependence on mutual trust, increased use of computer technology, greater flexibility in organisational processes and, ultimately, uncertainties with regard to jobs and tasks. We take the term "corporate culture" to be an elusive concept that focuses on capturing "the essence of people". Only corporate culture does justice to the image of the "complex human being" in its variety and diversity of human behaviour, which also relates to the complex of decision-making processes. Decisive for the quality of a decision are the decision premises, the readiness of the decision recipients to carry out the decision and the decision time. This makes it understandable that corporate culture arises where problems arise that cannot be solved by existing instructions, for example in the case of disagreements. Corporate culture is only used in the singular, meaning that each company has its own culture, but one that is consistent within itself. Corporate culture refers to values shared within the company that do not have to be in harmony with social conditions. A corporate culture is created by itself through the way of internal communication in the company. Communication is seen as a three-stage process involving information, communication and understanding. In consciousness, something specific (other not) is noticed. This information is selected, translated into language and uttered. The communication is translated by another consciousness, processed and accepted or rejected by a subsequent action. Whoever feels addressed by the communication, his consciousness selects what he understands and how – and he then motivates himself, or refrains from doing so. This is accompanied by the fact that verifications of information are not possible in view of these communication channels. In order to be able to introduce doubts, queries or requests for justification, one has to participate in this communication process oneself. This is the only way to assess what information has actually flowed between decision-makers and decision-recipients and how it has been processed. Shared values also permeate the communication process. For the purposes of internal communication, corporate culture remains invisible (Luhmann 2006). Corporate culture can only be highlighted in comparison with other companies by emphasizing the specifics of one's own company. This is also a difficulty that regulators encounter when they are asked to assess procedures. The difficulties of conveying corporate culture without directly embedding people who should/will judge what is happening have also become visible in the three spectacular individual events.

At Chernobyl, no living safety culture was evident. The supervisory authority was not embedded in the entire test procedure.

The term corporate culture could be described as the typical Japanese management culture at Fukushima Daiichi. Safety culture in the true sense was only rudimentary. It was (and is) characterised by conscious acceptance of a known and existing risk. The role of the supervisory authority is not comprehensible.

In what happened on the Deepwater Horizon rig, a culture can only be described in terms of economic values. The regulatory authority "duplicated" the intended process instead of controlling it.

All three individual incidents show that corporate culture – i.e. safety culture – cannot be predefined because the specific intentions can only be made visible after the event. In all three cases, it was only after the accidents that it was determined how the respective safety culture was shaped. Or to put it another way: the break with the corporate culture only becomes visible as a decision taken by a dominant personality.

In Chernobyl, this was the load dispatcher, the person responsible for the electrical supply network in Ukraine.

The described safety status of the Fukushima Daiichi plant can only be attributed to management decisions.

This also applies to the operator of the Deepwater Horizon drilling platform, the US conglomerate Halliburton, whose incentive system did not (does not) focus on process safety.

Of course, corporate culture is not a recipe for success. It can only lead to this if executed decisions are questioned, especially with regard to the decision premises, and their results are fed in as "feedback" in order to change the existing preferences.

We would like to quote Luhmann (2006) on this:

> In organizations working with high-risk technologies, the main problem is to avoid situations with unclear, cognitively undefinable and above all: not quickly definable enough problems. For such cases, and for organizations where such risks are the main problem, the availability of complex and rarely (one hopes: never) used cognitive routines is the critical variable. One cannot rely on hierarchy on the assumption that better knowledge and responsibility are held at the ready at the top; nor can the system protect itself through normative programming, for it is clear anyway that disasters must be avoided. The problem is rather one of cognitive capacity: one must be able to remember something that has never happened.

The observation in the production of airbags mentioned at the beginning of this chapter fully confirms Luhmann's quotation.

The quotation from Luhmann and the observation described show that organisations must anchor competences where they are needed in the decision-making process. Communication channels that are fixed with the corporate hierarchy only have the purpose of transporting competence. The most important facet of corporate culture is the interaction of competence and communication.

Or formulated differently: In order to achieve short-term intentional goals, in the cases of the three catastrophic individual events the control loops stabilizing the work processes were broken up or even interconnected feedback loops were destroyed.

4.7.2   Decision Premises

In the cause-effect structure, decision premises form the preliminary phase. In the intentional structure, they form the reason for action (motivating intention). Decision premises are the tasks of an organisation, i.e. the operational purpose. Following the operational purpose, occurring errors must be hidden as far as possible. Therefore, self-reports describing errors of the company must be viewed with extreme caution. The policeman knows how to write his report! In this context, it is incomprehensible to pursue a corporate policy of absolute error-proofing. Management wants to set up a company so robust that errors do not leak out.

In the process, errors report themselves, so to speak. With target-oriented diagnoses and knowledge, one can find out how to eliminate them. In this way, errors become the impetus for the learning process, which leads to expanded transparency for the work process. It is inconceivable that all decisions of a company are based on the single overall operating purpose. Decision premises are discussed especially in terms of time in relation to risk. This term only captures consequences of decisions that can be attributed to decisions, i.e. would not occur if they had not been made (Hoensch 2006). These risk-influenced decisions, especially management's memory for them, are difficult to document or are not documented at all.

With these forms, the company achieves the seclusion that is necessary in order to concentrate on the purpose of the company and thus to be able to be entrepreneurially independent. Thus, the corporate culture can determine which factors are adopted from outside and incorporated as causal. Causality encompasses every company, if only because it is part of a "socio-technical system" (Hoensch 2006), keyword: consequential reasons for action.

Operational organizations are not machines, although both were created by humans. With machines, the consequences of interventions can be estimated on the basis of decisions. This obviously did not happen in the three individual events. In the case of interventions in organizations, the decision premises can be influenced by changing responsibilities, strengthening or weakening competencies in the hope that the interventions will be purposeful.

Thus, the companies interested in added value do not have direct access to the work performance as in a functional machine, but they have to reckon with the actions anchored in the personality system. Entrepreneurial organizations are not a productive cooperation, but a construct that depends on the cooperation of the employees, to which the employers have no direct access.

Management tries to increase the probability of occurrence for actions it desires. To do this, they lack direct possibilities of influence, since the employees form a biological-personal system that can only be influenced indirectly.

4.7.3   Decision-Making Processes

Corporate culture and decision-making processes are the two sides of the same coin "company organisation". The organisational structure is the totality of relatively stable patterns of behaviour that reduce the complexity of a work process to such an extent that it can be managed by decision-making processes.

In decision-making processes, a dual character also comes into play, as we already know it from actions, which is shaped on the one hand by the interaction of cause-effect structure with spatiotemporal behaviour and on the other hand by consequentialism. This dual character of actions is also shaped by the fact that human action is not a purely personal matter, but also a collective one, directed towards the corporate goal. Decisions also have a dual character, as we see in the "airbag" example. On the one hand, a decision serves the corporate goal of economic production, so it has a promoting character. On the other hand, a decision also has a negative momentum, as created by the decision in the production of airbags, as a result of which the user of a motor vehicle cannot assume a fully functional airbag in the event of a crash. Every decision thus has a favouring and a burdening character.

Following Schuler (1998), p. 426, we will tabulate the elements of the decision-making processes in Fig. 4.2.

Dimension	Rationale Decisions	Operational Organization	Corporate Culture/Culture
Goals, Object, Premises	Unique premises set by all parties involved can be shared. Alternatives are based on greatest benefit selected.	Unambiguous, hierarchical.	Ambiguous, precise and adaptive modifiable trouble-shooting.
Facts, Coherences	Largely with the corporate aim known.	The Corporate-management are the goals largely known allow but a discretion.	Organizationalstructures and decision-making processes are carried out by the employees lived and in whose consciousness simplified pictured.

Fig. 4.2   Decision models, based on the field research of Hoensch (2006)

Power and Control	Different leave premises do not uniform handle. The corporate top leads with help from corporate staffs members and line superiors.	Power concentration through aauthorized instructions.	Organizational hierarchy corresponds with problem hierarchy.
Decision-making processes	Dual character, beneficiaries and burdens at the same time, but largely orderly.	Not suitable for the troubleshooting or solution.	Distortion of the human Information-Processing in favour of own intentions.
Values, standards	Optimization, efficiency, economy	Experience, Order, stability	Overcoming individual rationality-restrictions.

Fig. 4.2 (continued)

In the model of rationality, it need not be assumed that the goals are shared by all; rather, they are enforced from the top of the organisation downwards by means of instructions and authorised programmes (see Chernobyl and Fukushima Daiichi). This formally rationalist way of exercising power is reinforced by the corporate organization. In contrast, corporate culture demands from management an unusual degree of information processing for employees. Individual decision-making and problem-solving behavior is referred to as "bounded rationality" due to the limited information processing capacity of humans. This term goes back to Herbert A. Simon, the father of modern problem-solving psychology. He was awarded the Nobel Prize in Economics in 1978 for his concepts. "Bounded rationality" manifests itself in the fact that problems are initially not recognized or denied, that the problem definition and objectives are only developed in the course of the problem-solving efforts, that they are based on a simplified picture of reality, that only a few alternatives are sought, that satisfactory solutions are sought instead of best solutions and that, if necessary, the level of ambition is lowered in order to arrive at a viable solution (Schuler 1998). Through corporate culture, the difficulties that organizational structures face in the decision-making process in dealing with problems are overcome through adaptive problem solving. On the other hand, values authorized by corporate culture can be overtly or covertly challenged and changed. All in all, corporate culture overcomes the limitations of rational decision making especially because safety culture/corporate culture cautions to be careful about declaring something as a fact. Once something is declared a fact, less attention is paid to what is happening, and attention to risk-conscious decision-making diminishes (Hoensch 2006).

Our field observation (Hoensch 2006) invalidates the argument that decisions can be regarded as a mystery. Only the part of the consequential reasons for action that we have identified as accompanying intentions remains in the dark. Which, however, contributed decisively to the three catastrophic individual events and was not, as in contrast to the ballad "The Sorcerer's Apprentice", compensated by a "Deus ex Machina", the unexpected helper from a predicament. How decisions "really" come about cannot be determined (Luhman 2006). We have found in the three individual catastrophic events that decision-making processes dominate. It is difficult to determine how information intake – manageable in the case of airbag manufacturing – influenced the decision. We can further state that the decisions in the three individual events were influenced by social or economic processes. Individuals did not independently contribute to the three disasters.

Thus, we have to deal more intensively with human errors in decision-making processes, i.e. in problem solving, going back to the field observations of Hoensch (2006). According to our understanding, errors occur when the motivating intentionality (reason for action) of the action does not lead to the desired result without an external event being able to be used for this.

Chernobyl The execution of the experiment ignored the thermohydraulic and neutron-physical behaviour of the RBMK reactor. The motivating action goal, proof that the rotational energy of the outgoing turbo set is sufficient for the aftercooling, could not be achieved, because the findings of the accompanying intentions (behaviour control), no comprehensive evaluation of the interim experimental results, remained unheeded.

Fukushima Daiichi The reactor was operated without the necessary precautions against predictable environmental events. The economic operation of the reactor plant, the motivating intentionality, was only given until the arrival of the tsunami wave at the reactor site. The accompanying intention, a reflective and systematic evaluation of the operating results, was not given the necessary attention.

Deepwater Horizon Time and deadline pressure led to a non-safety-oriented drilling technique. The intended crude oil production, the motivating reason for action, became a desire due to the explosion of the gas cloud above the drilling platform, because there was obviously no time for the accompanying behavioural control.

The "errors" occurred in the three individual cases as deviations from an arbitrarily defined target state (motivational intentionality). Very different settings led to the definition of the target states. Management and operational staff intervened at many points in the complex and dynamic situations in the three individual events to "save" the arbitrarily defined target state through preceding and accompanying intentionality. Our field observations confirm Schaub's findings; we quote:

> ... that rational planning and decision-making is an illusion. Often the information processing requirements exceed the problem solving capacity of the decision makers. They do not take into account the imprecision of the information that is available or obtainable, they do not take into account the effort that information gathering and analysis require, they do not take into account the difficulties of evaluation and assessment, they do not take into account that outcomes and values can change and influence each other, they do not take into account that people cannot include all possible actions and environmental factors at all, they do not take into account that in practice people need instructions for step-by-step action, they do not take into account that in reality there is a continuous stream of interrelated problems.

Or, in short: the decision-makers adapt their mental model to the event, or the decision-makers adapt the event, i.e. their perception of it, to their mental images.

This was expressed in Chernobyl by the "doggedness" with which the five-year decision was to be achieved and the non-observance of the intermediate results achieved in the test procedure.

In the case of the Fukushima Daiichi reactor plant, the management clearly gave preference to economic operation over knowledge of tectonic stress conditions. In addition, there was no self-reflection due to many years of operating success, thus confirming: success makes conservative.

On the drilling platform, there were multiple factors; analyses were omitted, attempts were made to solve only the current problems, due to "ad hocism" no action was thoroughly planned.

There are various psychological approaches that attempt to narrow down the area of causation of errors or error classification. It is important to distinguish between the frequency approach and the cause approach. In the frequency-oriented approach, the human being is considered as part of the overall human-machine system. In the cause-oriented approach, the psychological laws underlying the errors are to be inferred.

We will restrict ourselves to Reason's "generic error – modelling system" (GFMS), which builds on the work of Rasmussen. We have already described the work of Rasmussen and presented it as a "stepladder model".

It is precisely the typical level changes described by the "stepladder model", which are functional in the sense of problem solving, that give rise to problems in connection with human-machine interaction (Helfrich 1996). Possible sources of error lie in limited knowledge (keyword: gaps in knowledge), in the selectivity of observation, in the boundedness of thinking, in the tendency to ignore or explain away information that contradicts one's own opinion, and in the abandonment of critical examination (Helfrich 1996).

The error mechanisms in the GFMS are related to the three levels of execution already presented according to Rasmussen (1986), see Chap. 2, the skill-based, rule-based and knowledge-based levels. The functions of the skill-based level precede the discovery of a problem, the problem is followed by actions on the rule- and knowledge-based level.

We quote from:

> Problem-solving behavior according to the GFMS is based on the idea that people try to search for known patterns rather than optimizing elsewhere. Thus, they check their memory for rules that have been successful in similar situations and apply them before moving to the more elaborate knowledge-based level. Such rules take the form of IF-THEN statements. Only when the rule-based path fails to produce a solution is the switch made to the knowledge-based level. At this level, too, solutions are initially sought on the basis of memory content and cue stimuli. However, if there is a deviation from conditions in a known action during an attentional test, the system always switches to the rule-based level. When a suitable rule is found, the system switches back to the skill-based level. This cycle, may be repeated for difficult problems and/or inappropriate rules. The problem solver should switch from the rule-based to the knowledge-based level when he finds that no suitable solution is available to him. Once a suitable solution is found at the knowledge-based level, skill-based routines and courses of action are resorted to. Frequently, a quick switching back and forth between knowledge-based and skill-based levels is necessary. Due to the observable tendency to want to find a solution quickly, the problem solver will often accept inadequate or incomplete solutions.

It is precisely these level changes, which are typical for human actions and functional in the sense of problem solving, that raise problems in the context of human-machine interaction (Helfrich 1996). This aspect is hardly considered in frequency-oriented computational methods (Helfrich 1996), which is of enormous importance and unpredictable in knowledge-based actions.

This flexible action also complicates the analysis and evaluation of human action in risk studies. There, people are regarded as part of the system, as system components. He has to fulfil a certain task within a given time. If he fails to do so, the "human" component is considered to have failed (Hauptmanns et al. 1987), cf. Chap. 1. The reliability calculation is carried out separately for the technical and the human component with the aim of combining both probability values into an overall value for the work system and thus arriving at a risk assessment (Helfrich 1996). Attempts are made to get human errors "under control" by design improvements and automatic control devices. As a result, although errors at the capability-based level are decreasing, the proportion of knowledge-based errors is increasing. Knowledge-based actions are characterized by the fact that the error-triggering constellations of conditions cannot be predicted in detail in a concrete spatiotemporal behavior, as we have seen in the three catastrophic individual events. Compared to the technical components, humans are characterized by the greater variability and complexity described above. The description of his behaviour by reliability parameters should be treated with extreme scepticism. None of the known error models takes into account that the accomplishment of work tasks and the mastering of crisis situations are always accompanied by social conflicts.

Or to put it another way: Individuals do not contribute independently to the overall outcome, but are subject to interactions that lie in the social domain. This leads to a diffusion of responsibility across several individuals, which goes hand in hand with the diffusion of information.

Because of this problem, there is currently widespread agreement among experts that only actions or action elements can be adequately described by reliability parameters that can be assigned to the levels of skill-based and rule-based behavior (Hauptmanns et al. 1987), cf. Chap. 1.

In the following Sect. 4.7.4, we want to analyse the actions of humans in a self-dynamic, incomplete system that changes by itself without intervention by decision-makers. This requires some explanation. The cosmos can be regarded as a self-dynamic, unclosable system, i.e., the system is self-dynamic because it develops only according to the laws of cosmology, and unclosable because, despite its "openness", it cannot be "unlocked (influenced)" by human intervention. The laws of the cosmos that govern the life zone Earth are discussed in Sect. 4.8. Time pressure arises from the self-dynamics of the incomplete systems, which is intensified by the necessity to realise arbitrarily set goals. We do not understand time pressure here in a colloquial sense, but as a dynamic. For us, time pressure arises from the two arrows of time in the natural sciences, the thermodynamic and the cosmological. In the social-scientific realm through arbitrarily set goals, which we want to achieve with the psychological arrow of time.

4.7.4   Dynamics of Decision-Making

How time pressure arises in spatiotemporal behavior and thus influences the dynamics of decision-making can be illustrated by the physical quantities momentum p and energy E and explanatory examples, which merge into the concept of effect. Or to put it another way: The energy and momentum of a body indicate the entropic processes to which it has been subjected.

We want to explain the dynamics of decision-making in terms of entropic processes. Entropic processes are also understood as a thermodynamic arrow of time and as a cosmological arrow of time, which is regarded as determining the course of cosmic expansion.

In order to be able to apply the concept of entropy to social systems, we have to consider crucial prerequisites.

We would like to start with the definitional understanding.

Around 1880, Ludwig Boltzmann was able to explain entropy on a microscopic level with the statistical physics founded by him and James Maxwell. In statistical mechanics, the behaviour of macroscopic thermodynamic systems is explained by the microscopic behaviour of its components, i.e. elementary particles and systems composed of them, such as atoms and molecules. In short, the greater the entropy, the more indeterminate the microscopic state, the less information is available about the system.

In the perspective of natural science, the First Law of Thermodynamics (law of conservation of energy) and Boltz's Law of Increasing Entropy (Second Main Theorem of Thermodynamics) are framework conditions with validity up to cosmic dimensions, the utmost for the existence of living beings and thus social systems.

In a closed system, where there is no exchange of heat or matter with the environment, entropy cannot decrease according to the Second Main Theorem of Thermodynamics. The distinction between "open" and "closed" systems is important for the application of the entropy concept to social systems. Clausius (1870) said that the entropy of the universe is constantly increasing, leading to a maximum, although there are many subsystems in the closed system that reduce their own entropy increase by adding energy from other (open) systems.

As we have argued, entropy entered the social science field through information theory, according to which information can be understood as negentropy or entropy increase as information loss and thus responsibility diffusion. From a social science perspective, entropy is a "measure of disorder." However, disorder is not a physical concept and therefore, unlike physical entropy (joules per kelvin), has no physical measure. The social sciences usually use derived indicators such as sustainability for the metaphorical term "disorder".

Social systems, which also include economic enterprises, are, as we know from systems theory (Hoensch 2006), fundamentally open systems. Entropy reduction in an enterprise presupposes an export of entropy, i.e. services in the broadest sense, to the environment or an import of entropy in the form of information and knowledge from the environment. Open systems are shaped by decisions. Open systems can thus exhibit "microstates" of increasing and decreasing entropy according to the choices made. Power differences are instrumental to this exchange process among microstates. Despite this exchange of entropy, economic processes, business activities, production of goods and services cause an irreversible, i.e. without external intervention irreversible, increase in the entropy of the entire system, the entire universe. On this basis, sustainability therefore means keeping the entropy increase in the overall system as low as possible (Hochgerner 2012). Decision-makers must therefore pay attention to the "macro-states" of the overall system, the universe, as well as the dynamics of decision-making, starting from the "micro-states" of their own subsystem (company). They must clearly distinguish the impact of their decisions on the development of entropic processes in their subsystem from the impact of the overall increase in entropy. Decision-makers must avoid mixing the social and natural science domains, just as we have strictly distinguished between causes and reasons.

Or, to put it another way: physically, entropy is a thermodynamic state variable of the material environment of the economy and society. From a social science perspective, entropy is of interest primarily from the point of view of increases and decreases in entropy in fundamentally open systems (Hochgerner 2012). We will discuss the social science perspective of entropy again in the summary, Sect. 4.10.

Our economy is pure energy, material and social energy. The energy is not only in the raw or valuable materials, but also in the minds of the employees, the corporate goals, the desires within society. All energetic transformation processes create time pressure to act and lead to the question, how can entropic processes be controlled and thus counteract the time pressure?

For the answer we have to come back to the specific differences of social and natural science disciplines mentioned above.

Following the social science perspective, entropically relevant processes can be influenced by knowledge, consciousness and consequential reasons for action, but also by entrepreneurial decisions. The social science perspective thus allows the arbitrary setting of goals. It is of particular interest from the point of view of increases and decreases in entropy in open systems. A system can control the entropy increase inside in the area restricted to it and at the same time "export" entropy. The export of entropy leads to new system properties. Such system changes have certainly contributed to the fact that the world has always (still) been able to make enough energy available despite increasing energy hunger and exploitation of natural resources (Hochgerner 2012).

Export of entropy from social enterprises can be achieved by building new system structures.

In short, time pressure arises from the efforts of social systems to reduce the increase in entropy. However, the entropy of the system as a whole continues to increase. The entropy increase is intensified by power efforts aimed at maintaining existing system boundaries.

We would like to explain our understanding of the self-dynamic, uncompletable system mentioned above on the basis of the three catastrophic individual events.

Chernobyl   The focus of considerations of the operation of a nuclear power plant is normally on stationary, i.e. temporally unchanging energy production. However, the experiment at Chernobyl involved a reactor in a time-varying, i.e. dynamic, operating state (transients, i.e. transitional states). The treatment of reactor dynamics follows from the summary of reactor kinetics with the temperature coefficients and the equations for heat conduction and heat transfer determining the temperature field in the reactor core. The equation that generally describes the motion of particles interacting with others in various ways is called the Boltzmann equation or transport equation. It was introduced by us in the context of explaining entropy through the kinetic theory of gases. We therefore regard the reactor core as a self-dynamic system; as unfinishable; because the Point of no Return was exceeded with the shutdown of the last remaining safety system and the triggering event was created with the continuation of the experiment at 23:10.00 on 26 April 1986.

Fukushima Daiichi   We are getting a little ahead of Sect. 4.8. We now know that the thin crust of the Earth is not a closed shell. Far more than Earth-sized plates drift on the hot mantle below. The Earth barely cools inside, thus maintaining the drive of plate tectonics. For the Earth's heat balance, the energy in the Earth's interior is of little importance compared to solar radiation; however, it protects life in another, dual way: First, the melting of rock in the Earth's interior again ensures that the water and carbon dioxide bound in the rock are released; without this recycling, there would presumably have long been no more water and carbon dioxide in the atmosphere, and no life on Earth. And secondly, convection movements in the liquid outer part of the earth's core are probably the cause of the generation of the earth's magnetic field, this earth's magnetic field protects the earth from solar winds (In the north you can sometimes see these solar winds, aurora borealis).

The drive of plate tectonics is the self-dynamic part of the system. It ensures that a submarine earthquake (seaquake) is created by sudden lifting and lowering of parts of the seabed, which eludes any access of man and is therefore inconclusive (uninfluenceable). Unlockable, also because no precautionary measures were taken against tsunami waves, the Point of no Return was already set with the commissioning of the plant. The triggering event was the seaquake on 11 March 2011 at 14:46 local time off the coast of Honshu.

We would like to point out that an earthquake can only cause a tsunami if all three of the following conditions are present:

1. The quake reaches a magnitude of 7 and more.
2. Its hypocenter is located near the earth's surface at the bottom of the sea.
3. It causes a vertical displacement of the sea floor, which sets the overlying water column in motion.

The probabilistic drivers can also be quantified; only 1% of earthquakes between 1860 and 1948 caused measurable tsunamis.

Deepwater Horizon Most of the crude oil produced today is derived from dead marine microorganisms, with algae making up by far the largest part of the biomass. The biomass is chemically transformed with the help of bacteria, resulting in digested sludge. Petroleum cannot be extracted from this predominantly clayey petroleum source rock; it migrates into porous and fissured reservoir rocks, where it accumulates under impermeable cover layers in petroleum traps to form reservoirs. This is a self-dynamic process that cannot be completed (influenced) by humans. It is only through deep drilling, including underwater drilling, that oil deposits can be developed. At greater depths, the crude oil is under the pressure of the superimposed layers of earth and possibly the associated natural gas and is pressed out of the borehole after drilling, as it is lighter than water and the surrounding rock. When the reservoir is first drilled, the oil, which is under pressure (the reservoir pressure was approx. 900 bar), must therefore be prevented from escaping with the aid of a blowout preventer, which is known to have been inoperative. The Point of no Return was set by the wrong drilling technique. By replacing the drilling fluid above the cement, the drilling crew caused the triggering event.

We would like to return to the content of our anticipation, which also leads to system changes, the very greatest.

4.8   The Very Largest

At the beginning of Chap. 2, we quoted Dschung Dsi's question. We related the question to direction and reliability and examined the possibilities of influence on these two aspects by human decisions, as the very least. In this section, we would like to address the question of direction and reliability from the aspect, the Very Largest.

We have let ourselves be guided by the "Ariadne thread", which is vital for us, and we have never let this thread depart. If this thread were to depart, it would mean running the risk of dissolving the network of knowledge built up through generations of human generations. We must unravel stitch by stitch in order to understand the network into which the arrow of time has forced us.

To do this, we start with the fact that our world has a beginning, a history. And it begins 13.7 billion years ago with the Big Bang (see Fig. 3.2, Chap. 3). For most scientists, this is the zero point that cannot be explored in principle. At which both space and time came into being. This unexplorable period of time is called the Planck era. A single power determined the events in the just born cosmos: the primordial force. It directed the events that by chance, according to the speculations of cosmologists, triggered a spontaneous fluctuation of energy and thus the dramatic explosion. The expansion – and thus the cooling – of the cosmos caused gravity to split off from the primordial force: the force that acts as an attraction between two massive bodies and causes planets to revolve around stars today. Gravity resists the pressure of the universe. But it is not powerful enough to stop the expansion. Then the universe enters the next phase of development. Again, something random happens. What remains of the primordial force decays into two further forces: the strong nuclear force, or nuclear force for short, which holds atomic nuclei together despite the positive charge of the nuclear particles, and the electroweak force (a predecessor of those forces whose action, according to present-day knowledge, is responsible for light radiation, electric current or radioactivity). Now something begins which cosmologists call inflation, i.e. gravity no longer slows down the expansion of space. On the contrary, it even accelerates its expansion. The electroweak force further divided: into the Weak Nuclear Force, which makes neutrons decay and causes radioactivity (without the Weak Nuclear Force, according to the laws of physics, matter and antimatter would annihilate), and the Electromagnetic Force, which attracts differently charged particles and makes negatively charged electrons buzz around positively charged atomic nuclei. According to the electroweak theory, two fundamental forces – electromagnetism and the weak force responsible for radioactive nuclear decay – act differently today, although they once formed a single, unifying force. The Weak Force and the Electromagnetic Force began to behave differently: The symmetry that had previously unified them was broken.

From the original primordial force, those four forces emerged, gravitation, electromagnetic force, weak and strong nuclear force, which control all processes in the cosmos until today. All four forces are exactly "balanced" so that our material existence is possible. Smallest variations would destroy the physical world.

Why was it possible for higher life to develop on Earth?

How "balanced" our life on earth is, shall be described by the example of the "fusion reactor" sun.

The thermonuclear-fed radiation at the surface 6000 °C hot plasma sphere is the basic prerequisite for the emergence and development of life on Earth. To compensate for the entropic processes on Earth, the Sun itself loses four million tonnes of mass per second in the form of radiant energy. Nevertheless, it can probably radiate energy for 10 billion years and in the end it will have lost only about 1% of its original mass. The conversion of three solar masses in less than a second thus provides an incredibly large amount of energy. Despite the enormous energy conversion, presumably no signal could be observed in the visible light range or in any other part of the electromagnetic spectrum. The unimaginable amounts of energy were radiated by gravitational waves alone. The solar system was formed 4.6 billion years ago by the gravitational collapse of an interstellar gas cloud (star formation). The subsequent evolutionary history of the Sun leads to its present state (yellow dwarf). A yellow dwarf stays in the main sequence for about 10 billion years during its existence. In the course of its existence, the yellow dwarf evolves into a red giant and finally, via an unstable final phase at the age of 12.5 billion years, into a white dwarf surrounded by a planetary nebula (Spanner 2016).

Who directed this process?

For the complete answer, other parameters for the prerequisites of life on Earth are exemplified:

The correct position in the Milky Way. The correct position in the solar system. The role of the atmosphere. The earth as a system. The help of Jupiter and the moon. Is there extraterrestrial life in the universe?

For the question of direction, let us look at the parameter, the earth as a system.

What might the "thermostat", which we have already addressed by setting a set point, of the earlier Earth have looked like? (Thermostat in quotation marks, since there is no set point that anyone has set, but there are control circuits that keep the earth's temperature within a range that is conducive to life). This requires control loops that lower the temperature when it rises, and others that raise it when it cools. Such "system-stabilizing" control loops are also called negative feedback loops (negative here means that they counteract changes). In addition, there are also positive feedbacks that amplify the effects of changes, and which therefore determine the inner and outer limits of the living zone.

The control loop that determines the inner boundary of the life zone is the water vapor control loop: Rising temperatures (due to the planet's approach to the sun or due to a hotter sun) lead to greater evaporation, and this leads to a higher water vapor concentration in the atmosphere. Since water vapor is a greenhouse gas, this further increases the temperature, which in turn increases the water vapor concentration. This is a positive feedback. However, this is slowed down by an effect on which the functional principle of pressure cookers is based – as the water vapour concentration rises, additional evaporation becomes more difficult. The area in which the temperature is not yet high enough for this is the inner limit of the life zone.

The outer limit of the life zone is determined by the ice-albedo control loop (albedo, from Latin albus = white). As the planet moves further away from the Sun and cools, ice caps form at the poles and expand with further cooling. This is also a positive feedback and beyond the outer limit of the life zone, this feedback circle causes all water on the planet to freeze and no more liquid water to occur.

Negative control loops stabilize the life zone by making the planet less susceptible to changes in, say, solar radiation.

Result: These control circuits ensure that the inner and outer limits of the life zone cannot be reached so easily, thus stabilizing the planet in the life zone.

More generally: control loops are based on feedback processes, thus control loops maintain order and promote stability. Not only in the "control loops life zone earth" but in all control loops an actual value-setpoint comparison constantly takes place, in which the inputs can no longer be clearly separated from the outputs and thus causes from the effects. This also applies to the brain. Thermodynamic constraints cause our brain to regulate the "chaos" caused by disturbance variables to stable, i.e. ordered, states. This happens via synapse weighting (synapse: contact, switching point between nerve processes, at which nervous stimuli are passed on from one neuron to another; an estimated 200 to 300 bio. Synapses in the brain are connected to each other. Synapses in the brain trigger so-called action potentials, then signal chains) and the strength of the connection through a finely tuned interplay of activating and inhibiting circuits. Evolution has ensured that only those circuits prevail that provide a picture of reality in which humans can find their way.

On earth, however, it should not remain with the inanimate control circuits, with which actually the question of Dschung Dsi would be answered. These control circuits have made the earth a system – in a system the elements interact with each other in such a way that they can be regarded as a unit. On Earth, life evolved early. The age of our solar system is thought to be about 4.6 billion years. The oldest traces of life known so far have an age of 3.85 billion years. This means that our Earth may have remained uninhabited for just under 800 million years, Chap. 2 Penzlin (2014); (Franken 2007).

But man began to interact with these inanimate control circuits of the earth: The system became the Earth ecosystem.

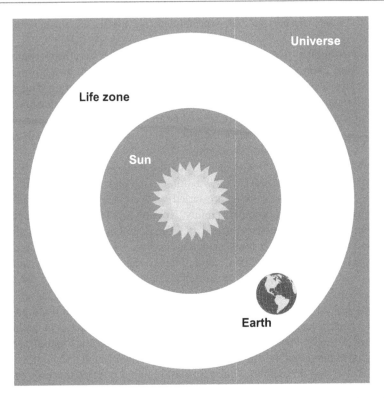

Fig. 4.3 A prerequisite for life on Earth: Our Earth is located in the life zone of our Sun – in an area where solar radiation is neither too strong nor too weak for life to thrive

Example

At the end of the nineteenth century, the Swedish chemist Svante Arrhenius realised that the carbon dioxide content of the air influences the temperature – the more carbon dioxide in the air, the warmer the earth becomes. The sun was formed at about the same time as the earth, and at the beginning only gave off about 70% of its present radiation. So the life zone was – see Fig. 4.3 – much closer to the Sun in the early days of the solar system. And yet geological studies and also the history of life show that for over 4 billion years the temperature of the Earth has fluctuated by only a few degrees Celsius. Humans have broken the inanimate control loops through their activities. This example shows that human activities, which are summarized under the term climate change, fall into the responsibility and thus the direction of humans.

According to the principle of causality in physics, however, every effect has a cause. Therefore, new attempts are constantly being made to find the deeper reason for the accelerated expansion (cf. Fig. 3.2, Chap. 3) of the cosmos. All approaches to combine gravity with the strong and electroweak interactions led to fundamental difficulties (Spanner 2016). We quote Spanner: "To date, physicists have not succeeded in formulating a widely accepted, consistent theory that combines general relativity and quantum mechanics. The incompatibility of the two theories therefore remains a fundamental problem in the field of theoretical physics" (Spanner 2016). If we succeeded in detecting a trace of cosmic inflation, our empirical knowledge of the cosmos could grow in unimaginable ways. We could then establish laws for the time evolution of elementary particles and the way to derive probability statements about experimental results from these laws (Spanner 2016).

"We could think of developing a universal law of nature that also allows an answer to the following question: 'What time evolution of a system accessible to us can we typically expect under ignorance of the exact initial conditions'" (Spanner 2016).

An all-encompassing theory is still the ambitious goal of modern physics, which includes quantum mechanics and relativity as limiting cases. This would also be in the spirit of Einstein (Spanner 2016). A particular challenge for a comprehensive physical model, however, remains cosmology and the explanation of its hitherto unexplored phenomena (gaps in knowledge).

Just as we have established on the basis of chaos theory that a flap of a butterfly's wings can cause typhoons because of an extreme sensitivity to slight changes in boundary conditions, the same applies to cosmic inflation (see Fig. 3.2 lower part, Chap. 3). Until then, for the question of direction, there remains only the reference to Lesch (2016), Chap. 2, according to which the four elementary forces can be considered for this purpose. And for the probability statements the "tunnel effect" introduced by Lesch. Or the quote in (Spanner 2016) by Danzmann: "… you just have to run with your head against the wall – until it collapses."

These two lines of thought, although empirically without support, presently provide the only naturalistic explanation for the occurrence of extremely improbable cosmic initial and boundary conditions. The model is the explanation of the suitability of the physical boundary conditions on Earth for the origin of life. The moderate temperatures, the stabilization of the earth's axis by the moon, the formation of liquid water, the earth's magnetic field and the ozone layer with its shielding of the earth's surface from the deadly radiation of space provide excellent general conditions for the emergence of life. This fine-tuning is made understandable by the fact that there are a myriad of other planets that lack these.

It follows that, on the basis of our present knowledge, we ourselves must take care of the shaping of human life and social coexistence, that is, we must use our freedom of will and self-realization responsibly. This thought was formulated by Napoleon Bonaparte as follows: "To do one's utmost in this best world, and to find one's reward in one's own consciousness, that is the great secret of never becoming a cheat or a flatterer, never bitter, troublesome, or a criminal."

On the question of directing: We created a stage for ourselves in which we can experience ourselves. The foundations of this stage are time and space. The direction lies with the four elementary forces and the script is written by the laws of nature.

We quote Bräuer (2005), see Chap. 3, and thus go back to his reflections on how evil came into the world, which we addressed in Chap. 3:

> Humans created an image of themselves in space and time, and this image continued to create, and so a world full of fractal (manifold) structures evolved.
>
> There must have been a problem at some point. In space and time nothing can be perfect, space and time are filled with excerpts (excerpts, note by the author) which cannot be completely compatible with each other. The image that people had developed of themselves in the world thus created could not fit the inner image they had of themselves. The need for projection arose. People identified themselves with the acceptable aspects of this spatiotemporal development, and they projected unwilling things away from themselves, into nature, which they have not yet succeeded in mastering completely, and into their fellow human beings. And this is probably how evil came into the world, as a projection of one's own archetypal (according to the original form, author's note) shadow.
>
> More and more the world was experienced separately as something inner and outer. And this process of separation lasted for millennia and only found its conclusion in the Middle Ages in the so-called Cartesian cut, in an absolute separation of the inner life of the soul and the outer, material world. In the outer world more and more laws of physics manifested themselves.

Regarding the term "Cartesian cut": It goes back to René Descartes (1596–1650). He was a metaphysical (transcending any possible experience) dualist and no longer separated the world into non-organic and organic, as was common before him, but into the world of extended bodies and that of mind and thought. Science henceforth took up the objective world of bodies, philosophy the subjective world of "thought." This "Cartesian cut" paved the way for science to single-mindedly pursue its empirical research without having to reflect on God and ourselves at the same time (Penzlin 2014), cf. Chap. 2.

We would like to relate the Cartesian cut to our thoughts on causality and the consequential reasons for action that act on causality by means of arbitrarily set goals, as well as to Kurt Bräuer's thoughts on space and time. The image that people created/create of themselves cannot be created by means of ideas of space and time alone. According to the Cartesian cut, we experience our world separately as a spatially and temporally extended physical world as well as the world of our sense impressions. The world is experienced separately as a subjective inner world and an objective outer world, the external. Our sense impressions must have a cause that is not within us. We juxtapose the images of the past, present, and future, and thus actually experience time in a spatial ordering structure. Time is not something objective. Time is objectified through the measurement of time. With coordinate systems we make images of temporal references. This time is a cultural achievement of mankind. Our existence, on the other hand, is based on a perpetual, ongoing present (Penzlin 2014), see Chap. 2.

We trace our sense impressions back to perceptions. We call the traceability to a cause causality.

We will illustrate these considerations in Fig. 4.4 as follows:

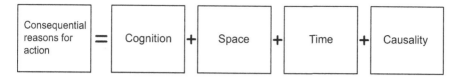

Transferred to the three catastrophic individual events, the addition is, whereby we fall back on the previous explanations and work only with concise terms.

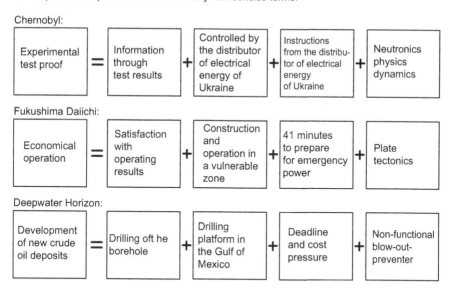

Fig. 4.4 Graphical summary for the three individual events

After transferring this to the three catastrophic individual events, here is our basic consideration once again: When we receive information, we immediately interpret this unconsciously spatially, temporally and causally in a cognitive process. Following this interpretation, we form consistent reasons for action for the setting of arbitrary goals.

Or, to put it another way, we assign a specific cause to the information. Only when we have identified the information temporally, spatially and causally do we begin to form consistent reasons for action.

As we have argued at length, space, time and causality are physical abstractions. Space, time and causality are, moreover, the three "vectors" that span our reality. And this reality, the fit of the Very Smallest in the Very Largest, is great and inexhaustible, the foundations of which man should not shake.

Man experiences an inner and an outer world. The inner world is shaped by individual and sociological circumstances. The external world manifests itself in the laws of physics.

We join Bräuer in Chap. 3:

> We live in a world that is essentially determined by the way we think. Physics has played an essential role in the development of our systems of thought and imagination. Mathematics and physics have led the way to a way of thinking that recognizes and explains more and more details. In the last few thousand years, mankind has thus come into possession of an immense treasure of knowledge. (Bräuer 2005)

However, our knowledge remains incomplete, as we have pointed out several times. We do not understand many important connections, there are still gaps in our knowledge. We have had to demonstrate and therefore understand this by working through the three catastrophic individual events.

We are still missing an important building block for the fit of the Very Smallest in the Very Largest. It relates to the concept of "causal intervention".

We have already addressed it with the introduction of consequentialism. As a reminder: Consequential reasons for action are directed at causally intervening in the world and thus bringing about a state of affairs that is different from alternative states of affairs as a result of the action, cf. on this Chap. 2 (Nida-Rümelin et al. 2012).

We want to focus on reasons for action that use the energetic processes to realize alternative states.

4.9   Energy Development of Society

We have established that our economy is pure energy, both material and social. This also applies to the phenomenon of "life". Energy is permanently required for the sole reason of compensating for its own decay, of maintaining the living state against the destructive forces that would lead life into thermodynamic equilibrium (finiteness). Organisms need energy even when they are not doing any external work (Penzlin 2014), Chap. 2.

With every transformation of energy from one form into another, neither energy is lost nor does new energy arise. The law of conservation of energy is always valid, sociology can not escape this approach.

The understanding of energy in the social sciences and natural sciences differs, as does the understanding of physical and social entropy, as we have shown.

In the social sphere, energy means a moral quality with which a person tries to fulfil a task set for him with strength and perseverance, despite possible resistance.

For natural science, energy is the work stored in a physical system, the work capacity.

Two images of energy that do not seem far apart in terms of language.

The physical law of the conservation of energy allows a balancing which enables us to make a statement about every natural process and to understand it, perhaps even to control it. In contrast, the application of the law of conservation of energy in the social sphere is largely unknown, although it would certainly be not only useful but necessary to make certain statements from this law for the social sphere as well. Let us only think of the introduction of gunpowder, chemical energy, as a result of which castles and walls for the protection of social communities lost their significance, and the formation of standing armies became a necessity (Ostwald 1909). Gunpowder was a new source of energy, much greater and more intense than the energies of human and animal muscle previously available for war. More generally, man replaced the causal chain of muscle power to suppress the enemy with a new form of energy. This led to a new social order. All events evolve from energy transformation. Without energy transformation, nothing would ever happen. Our quest is to increase the amount of raw energy (Deepwater Horizon) and improve the conversion to useful energy (Chernobyl and Fukushima Daichii). We briefly touched on improving useful energy in the archer through the planned use of drones. In energy conversion, a distinction must be made between the irreversible or quiescent part, which cannot be put into energetic conversion but is always increasing, and the reversible or moving part, which is responsible for what happens in the world. The Second Main Theorem of Thermodynamics states that the portion of free energy can only decrease, be consumed, but never increase. Because of this, and because of our finite nature, time itself enters our world of experience. We consume free energy in order to gain time (see Fig. 3.5). We have here an example of considerations underlying consequential reasons for action that lead to achieving the economically best result. Mathematically, the search for economic success presents itself as the solution of a maximum task. In the social sphere, this task is solved by decisions. Goal-oriented decisions are developed further, negative decisions are analysed. Thus, the Second Main Theorem of Thermodynamics becomes a guide for the development of mankind. The cooling process of the earth was already so far advanced at the appearance of the first human beings that the earth practically no longer contributes to the energy needs of its inhabitants. Only the sun constantly sends us energy, and free energy at that, and at the expense of this free energy practically everything happens on this earth. Only the tides and the phenomena that depend on them do not come from the radiant energy of the sun. A sustainable economy must be based exclusively on the regular exploitation of solar energy (Ostwald 1909). The energy released by the sun through nuclear fusion and radiated from it leads to radiant energy, a state of free energy, through an increase in entropy. With the exception of nuclear energy and the tidal energies caused by planetary motion (gravitational force between the earth and the moon), almost all of the forms of energy available for human use are based on solar energy. This applies primarily to solar radiation, but also to water power (solar energy acts as a motor for the terrestrial water cycle via evaporation processes), the potential uses of biogenic energy sources (solar energy is the driving force behind the growth process of plants), wind energy (the different heating of air layers leads to pressure differences and thus ultimately to the formation of winds) and the resulting wave motion and ocean currents. Last but not least, it is also the origin of the fossil fuels available to mankind, i.e. coal, oil and natural gas, which have formed over many millions of years from dead biomass and today represent the main pillar of energy supply with a share of around 80%, see Chap. 3. These are examples of the indirect use of solar energy. Direct use is represented by solar cells and solar thermal heat generation, even if they only succeed to a very small extent.

We also find this competition for free energy in biology. A look at the forest shows us how the assimilation organs, the green leaves of each individual plant, arrange themselves in such a way that they receive the greatest possible amount of solar energy. Each plant strives to get its share of the sun's radiation. Competition thus occurs for the limited amounts of energy. This competition extends to fauna and flora, but means nothing more than better utilization of free energy through adaptation. Man, on the other hand, does not regulate his relationship with his environment by adaptation, but protects himself by shaping his immediate environment so that it corresponds to his desired condition (keyword: building a house). He could have built the Fukushima Daichii plant to withstand a seaquake (tsunami). An animal cannot take such precautions.

Man also entered into competition with himself. He replaced the throwing spear with the bow and arrow. He used the elasticity or the form energy of the bow. The form energy could not be used with the rigidity of the throwing spear. Now the muscular work was first transferred into form energy and then in the next step into the kinetic energy of the arrow. So there was a temporal separation of the work effort of the muscles for the bow tension and the targeted relief for the launch of the arrow. This made aiming more accurate than with the throwing spear. The gain in accuracy of aiming also improved the use of energy, see also our reference to the use of drones to improve aiming accuracy. The use of the bow and arrow ultimately means a refinement or increase in the value of the energy used. If the throwing spear failed to hit, all the energy expended was in vain; the bow and arrow improved the ratio of total energy to useful energy. This ratio was further optimized by the use of chemical energy (poison in the arrowhead). Herewith we have gained a first overview of the energetic beginnings of culture. This resulted in an incomparably greater freedom for man and a substantial improvement in his food intake. What is remarkable about this is that the same or similar processes took place independently of each other in different places on earth.

Another, much more far-reaching superiority lies in the fact that man understands how to fulfil his purposes also by means of such energies which do not originate from his body but are taken from the outside world. This was the decisive step towards the dominion he sought over (Ostwald 1909). The first step was to consider animals, in addition to the intake of food from plants, as sources of energy, first for food purposes, but also as protection against the inclemency of the weather (use of fur), then for the performance of mechanical work. Then followed slavery in the use of foreign human energy.

Very similar considerations are possible for plants. Starting with the seasonal use as food and continuing through their cultivation as well as their energetic use, for example as biomass.

The social factor in the energetic use of animals and plants comes into play in that the knowledge of the use of both cannot be the achievement of one individual, but requires the joint work of numerous individuals over several generations. More generally, humans have been able to improve their own energy balance by procuring energy from outside. In animals, on the other hand, it has remained unchanged. Consider, for example, the tremendous efforts and dangers that migratory birds take upon themselves to escape winter and food shortages because they cannot improve their energy balance. In animals, nature balances the existing opposites. Man has not only created the balance of possibilities for his existence against nature, but in addition has improved his living conditions, in which he believed to have found a confirmation of his desired dominion over the earth.

Through this development, man has also exposed himself to the time pressure mentioned, which is created by his efforts to slow down the increase of entropy in a specific system. This is also his shaping for the Very Smallest. He succeeded in increasing the zones of life on Earth. But only within the framework of the life-zones, which the Very Largest (see Fig. 4.3) has "assigned" to him in the cosmos.

The historical development in the interpretation of the Second Main Theorem of Thermodynamics means that with each historical stage of development the supply of available energy decreases, while the disorder within the social and economic as well as the ecological structures increases (Suppes 1984).

Or put another way: The economic principle and the entropy principle are two sides of the same coin.

In his activities, man takes no account of the astrophysical constitution of the Earth, which determines the thermodynamic periodicity of the day and the year in terms of light and heat. Man makes himself free from the impediments imposed on him by the change of day and season. We quote Ostwald (1909):

> It is precisely this breaking free from the "natural" order of things that has found passionate expression in the Prometheus saga as a revolutionary mood against the "gods", i.e. the representatives of nature not yet influenced by man.

This mobility of man naturally also has repercussions on the transport possibilities of the forms of energy. Electrical energy has prevailed because it can be used almost without delay, bound to lines. Despite this mobility, man, like any other living being, remains bound as matter to space and time. This binding can only be dissolved by equivalent energy, which man supplies himself with through metabolism. Free energy decreases by itself with time. Hence our problems in dealing with time and space, because life requires a spatial and a temporal accessibility to the required energies. The most efficient way to achieve this accessibility is through collective purposeful action by people, which requires the coordination of the energetic trans-formation and the people involved, which in turn is controlled by the power of the decision makers. Thus our considerations lead to the principles of control.

Of the substances that are available for free, in the sense of common property of all people, use, the oxygen of the air belongs first and foremost. Without it, the use of chemical energy, food and fuel could not be realized. The oxygen of the air is therefore a form of energy that is not subject to private property, just like solar energy. We must distinguish between solar heat, which is freely usable, and solar radiation. Solar radiation becomes private property, in the form of photovoltaics, but also in the form of agricultural land. The same applies to hydroelectric power and mineral resources of all kinds. Solar heat can be used directly. Solar radiation only indirectly, it becomes a means of production with a "modest" efficiency in the sense of economic doctrines. "All transformation of raw energy into useful energy, by which human culture differs from the immediate life of animals and plants, presupposes human activity or human labor." (Bräuer 2009), says Bräuer.

In other words: Every finished product contains a proportion of energy as material and a proportion of energy that had to be expended to produce it and no longer exists as free energy. The sum of the two can be considered the manufacturing value of the product. The manufacturing value can be increased by stockpiling by the owner of the product. The owner can increase the value by exploiting space and time by offering the product to the potential user directly (space) and as needed (time). Thus, from an energetic point of view, "evil" enters the world, reinforced by the fact that the various stakeholders join together to form social communities (associations) in order to be able to influence future events with a greater or lesser degree of probability.

And why do we want to influence (foresee) the future? Because we want to control our actions in such a way that we want to achieve the desired alternative state with the least amount of energy.

In order to achieve the control of action goals, we need knowledge of the laws of nature, a science that stands for the fit of the Very Smallest in the Very Largest, and which is also the guiding framework for our social development.

The linking of social influences with the laws of thermodynamics is the central element for us, because we regard the social influences as the decisive momentum for the decision-making process and thermodynamics sets the "guard rails" for this. For this purpose, we refer back to a lecture that Boltzmann gave to the Vienna Academy of Sciences in 1866. In it he places the statistical approach, which he introduced into the natural sciences, in a context of demographic and social statistical surveys. "If external conditions are approximately stable, the number of voluntary and random social events ... calculated on the population remains constant." cf. (Müller 1996) in Chap. 2. He continues, "It is no different with molecules." (Müller 1996) in Chap. 2. Even then, the associations attached to the concept of entropy expanded, that is, into further areas such as cosmology, which stands for the Very Largest, and human coexistence, which stands for the Very Smallest.

The general struggle for existence of living beings is therefore not a struggle for the basic substances – the basic substances of all organisms are abundantly available in air, water and earth –, also not only energy, which is unfortunately abundantly contained in every body in the form of heat, but a struggle for entropy, which becomes disposable through the transition of energy from the hot sun to the cold earth. (Müller 1996), Chap. 2

It therefore does not seem unreasonable to seek analogies between the laws of thermodynamics and the laws of the social sciences. As Hochgerner (2012) and Brunner (1997) have done, for example. While in Hochgerner's case a regret shines through that this has not been comprehensively achieved to date, Brunner comments on entropy becoming a "world formula".

For us, social order corresponds to a high degree of social control and firm normative bonds.

Or put another way: The lower the social cohesion, the greater the social entropy.

Science in the sense we understand it naturally also draws attention to the gaps in knowledge referred to. These gaps in knowledge become the subject of further research and lead to an organisation that not only closes the gaps but also saves energy. A scientific network is created, such as is available to us today through the Internet. This network also achieves that various scientific images are revealed that need to be resolved in order to move humanity forward and limit the inevitable diffusion of responsibility. Science as the very greatest leads to the development of social traits or in other words the formation of character.

In short, knowledge must address the gaps in our knowledge of nature and shape character development through self-realization and free will.

We have dealt with entropic processes and processes initiated by consequential reasons for action. The two processes differ in that entropic processes occur without human activity. Consequential reasons for action, on the other hand, are affected by information processing and the decision-making processes. They focus in particular on social influences in the preliminary phase of the cause-effect structure.

We consequently expand our world view.

4.10   Our Expanded World View

Before we describe our extended world view, we would like to present our basic considerations. We orientate ourselves on Kurt Bräuer: "The world view of modern physics" (Bräuer 2009).

As a metaphor, we use the rainbow, as does Bräuer. A rainbow appears when sunlight shines on a wall of rain. The position of the rainbow depends on the position of the observer. The rainbow itself is a phenomenon, not an object. Without observation, there is no rainbow. In this picture, human existence is based on a big bang, a cosmological development and biological evolution. From which, in the end, man is supposed to have emerged. A relativization of this classical world view is certainly significant. The rainbow can help us to understand the most important aspects of modern physics (Bräuer 2009). Space and time are relative in the sense in which the rainbow is also relative. We experience space, time and matter as an external world and as the basis of our existence. Just like the rainbow, atoms exist only in the context of concrete observation. This cannot be grasped. We can only comprehend material world contents. We cannot grasp the rainbow either. If we approach it, it recedes and disappears in the end (Bräuer 2009). We can only understand the logical connections. And we can understand them in such a way that our view of the world is precisely a picture of the world and not the world itself. Between cause and observation, effect develops as a superposition of possibilities. Only in observation does the real, unambiguous world manifest itself, which we become aware of. Without conscious observation, this reality is inconceivable (Bräuer 2009).

If one transfers these considerations to the various aspects of the cause-effect structure mentioned, in particular causality, spatiotemporal behaviour, consequential reasons for action, time perception, right up to the problem of consciousness, then much can be explained by the fact that our organism has developed in such a way that we can cope in the environment described by the known physical laws. So the goal is not to get an objective (whatever that is) knowledge of the "true" world, but an interpretation that allows us to live a reasonably safe life, although possible dangers and risks can by no means be excluded.

The insight is: Nature is essentially only concerned with the preservation of the species or, even more generally, with providing mechanisms that ensure the accelerated increase in entropy. Nature has no regard for the welfare of the individual. In today's fixation on the individual, one may well see something "evil." A glance at history confirms our observation. We can recognize with horror that there have been and still are power structures which did not/do not care at all about the well-being of the individual – especially of the anonymous individual.

We have seen, especially in the events of Chernobyl and Fukushima Daichii, that when people's individuality is suppressed, their reaction is to submit to the guidelines and withdraw from conscious participation. The situation is somewhat different with the Deepwater Horizon drilling platform. Here, deadline and cost pressures led to peer pressure that resulted in the non-safety oriented actions of the employees. This behavior was reinforced by the passive role of the regulatory agency. We can speak in the three individual events of a resigned surrender to the given circumstances in which the individual perishes. The individual is traded as "capital" to be used in the interests of the system. The parts of society now exist separately, and a worldview of opposites exists (individual/social community, self-actualization/freedom of will, inner world/outer world, power/execution, possessions/needs, entropy/life, etc.). These fundamentals determine the rational structure of consciousness and, if they are followed blindly or if they are "idolized" as the only correct worldview (Chernobyl: Five-Year Plan), lead to a one-sided and inappropriate perception of the world.

In connection with our expanded world view, we must of course also address the "Newtonian world view", with which the model of reality, in that interactions between parts take place according to the cause-effect principle in a temporally linear sequence, contributes. Having migrated into the general consciousness, this world view is responsible for processes of social decay. An analytically one-sided science breaks down the physical world into mass points, organisms into cells, actions into reactions to stimuli, perceptions into sense data, evolutionary processes into the random mechanisms of natural selection. This is precisely the catastrophic horizon of the physical worldview, says Müller, Chap. 2 (Müller 1996).

However, we would also like to point out that we have established that causality can only be verified approximately and is limited to the boundaries of scientific knowledge. When these are exceeded by consequential reasons for action, the inadequacy of human beings becomes openly apparent, and the predictability of events is no longer given.

Our current worldview is materialistic and deterministic. When we receive a sense impression, we immediately and unconsciously interpret it spatially, temporally and causally – we assign it a spatial and temporally determined cause. We experience space, time and matter as absolute constraints to which our lives are subject, just as we are subject to sociological constraints. What counts in our world is what must be quantifiable.

Space, time and causality are not things or objects. Things and objects of all kinds are limited, finite and conditional. This is not true of space, time and causality. Rather, space, time, and causality are the three "vectors" that span our reality, the basis of all our cognition, the precondition of all objecthood. And because this reality is so large and inexhaustible, it can only be based on an infinite foundation.

Causality, as we have seen, finds its limits in relativity, in quantum mechanics, and in chaotic (determined random behavior) systems. Space and time are not the absolute worldly framework in which everything is calculable. What is observed depends essentially on the observation and somewhat more complex systems are usually incalculable, cf. Chap. 3 (Bräuer 2005).

The source of physics is our everyday experience of the world, and from this all physical details arise in a logically simple and unmysterious way. The properties of the physical forces and the fine-tuning of the natural constants result from the nature of our experience of the world, see Chap. 3 (Bräuer 2005).

The question of the direction of our actions answers itself, so to speak: Man directs with the consistent reasons for action that he has developed, which are formed from cognition, space and time as well as causality.

Man reliably follows the script of the laws of nature only if his deterministic worldview is in harmony with the purposes and goals of his action and probabilistics are well-disposed towards him. Decisions determine who has to appear when and where on the stage of events (space and time).

4.11 Safety Research

So far, safety research has not produced satisfactory results; on the contrary, see Charles Perrow, "Normal Disasters: The Inevitable Risks of Large-scale Technology" (Perrow 1987). The English title was even less provocatively: Normal Accidents (Luhman 2006).

This obscures a very deep-seated problem, society's irreversible dependence on technology, especially energy production, to compensate for entropic processes. Technology is a fixed coupling of causal elements. This coupling includes human behavior at the skill-based level, that is, unconscious behavior. This coupling can be separated by decisions at the rule-based and knowledge-based levels. The mechanization involves human perception (consciousness) and motor skills. Decision-making consists of finding alternatives among the available technical processes. The choice between different action alternatives is not to be seen as a random process, but is determined by factors of the situation, which are selected by the decision maker according to social norms (Hoensch 2006). Reasons for action can only be understood from the perspective of the actors. Actions are formed by actors from four elements: the subjective assessment of the situation encountered, the means available, the reasons for action and the social norms. Each decision maker faces a twofold problem of fulfilling his own intentions while taking into account the reactions of the people involved, as well as the causality of his action process. The problem here is that no sociological explanatory variables are available, since all organizational mechanisms are tailored to the psychology of individuals, which, as is particularly evident in the case of Deepwater Horizon, are based on gratifications for personal safety rather than sanctions for violations of system safety.

Modern society is more dependent on the provision of energy resources than any other society. As we have seen from the three individual events, decisions can lead to catastrophic consequences that not only affect humanity, but also the environment.

This degree of dependence on the supply of energy, which no one wanted and which has arisen evolutionarily, makes it clear what responsibility rests on the decision-makers. The fundamental question is therefore how society can influence organizations on whose functioning it is dependent, but cannot directly control them comprehensively enough. Technology and people are interdependent. Advancing technology creates an energy supply problem that can only be solved technically by man. Man creates organizations that can only be controlled by his decisions. This creates a mixture of technical order and human decision that can only be controlled by the influence of society (Luhman 2006).

The problem refers not only to the interaction of organizations and society, but also to organizations and the environment. We therefore speak of the social environment, which is not only characterized by unpredictable turbulence and non-transparent coincidences, but regards the social system as an ordering function. The ordering function is based on values, which are fixed points in the orientation of action, which must/can only be identified and demanded by the decision-makers. The decision-makers have to keep a special eye on the fluctuating economic parameters and adjust to the uncertainties reproduced in everyday life. The problem is, the uncertainties of the future that are constantly reproduced in the pace of work – see the manufacture of airbags – and the decision-makers must react to this in a timely manner in the decision-making processes in which they are embedded.

In short, man directs because he believes that he can master the tasks he has set for himself, and in this belief he displaces the rule of nature.

This is also supported by the studies of 100 shipping accidents cited by Helfrich (1996). According to these, only 4% of accidents occur without human intervention, i.e. are attributable to technical failure. In contrast, 96% of accidents are due to human acts or omissions (Helfrich 1996). Helfrich further examines social influence in hazardous situations in more detail in (Helfrich 1996). From her considerations it can be deduced that social influence is particularly strong when two people are involved between whom there is a power and status differential. Another influence highlighted is the social influence on risk acceptance (Helfrich 1996). According to this, socially shared norms often include precisely the tendency to circumvent safety regulations or not even to take them seriously. Social influence thus favours carelessness because non-compliance with regulations is already implicitly seen as a mark of "experience". Even more frequent than the errors of carelessness, which would have to be located on the level of rule-based behaviour, are the errors on the knowledge-based level (51%, Helfrich 1996), because unexpected events were perceived but not recognised as threatening. Hoensch's (2006) field observations also lead to this assessment. It takes a high degree of professional and social competence for the status low to at least doubt the decision maker's decision. This conflict hinders attention to the actual conflict situation and ultimately leads to the Point of no Return not being recognized.

As a consequence of the improvement of human direction, the proposed organisational measure remains for a lived safety culture, which leads in particular to a change in decision-making behaviour in risky situations between decision-makers and those of lower status.

This proposal needs to be supplemented with regard to the Fukushima Daiichi disaster triggered by the tsunami. The causality consisted in the tectonic movement of the earth's crust, which also established the temporal relationships. And this time factor is not always clear. For example, it cannot be said when the falling of the barometer leads to an approaching storm. The only statement that can be made is: It is necessary to take precautions. Not to take precautions is criminal.

4.12 Summary

Man is part of nature in all dimensions of his existence. As a person he belongs to nature like all material things and animals of this world. His creative power is solely the result of an evolution that has progressed over thousands of years, beginning with the earth that was formerly devoid of human beings.

Human action can be described and explained in terms of the cause-effect structure and the intentional structure in the spatiotemporal relational framework, including communicative influences.

The awareness of principally unpredictable events, such as the three individual events presented, the influence of decision-makers and the temporal dimension of irreversible processes demand, as it were, a dialogical attitude towards nature, or as Ilya Prigogine (1977 Nobel Prize in Chemistry) puts it: "respect, not control".

Man can develop with nature and his intentions in the sense of a prosperous development for both, under the limitations that the three individual events have shown.

Future security research should provide impulses for how a unity of their work can be established in the future.

The unbroken specialization within existing disciplines and the hardly manageable emergence of new disciplines are interpreted as an irreducible (not derivable) divergence between empirical and theoretical science and between the disciplines, which condemns all interdisciplinary counter-movements to failure (Suppes 1984).

What does this mean in concrete terms for the entropy concept advocated here?

We must strive for a unified understanding of thermodynamics as a branch of physics, which, starting from the study of heat phenomena, analyses all processes associated with energy conversions of various kinds and their applications. In this approach, depending on the system character and the methodological approach, a distinction is made between classical (phenomenological) thermodynamics, statistical thermodynamics and thermodynamics of irreversible processes.

The three main laws of thermodynamics are assigned to classical thermodynamics. The main laws of thermodynamics are formulated as postulates, which, however, are supported by all experiments:

The First Main Theorem of Thermodynamics (law of conservation of energy) includes heat as a special form of energy in the principle of conservation of energy, since mechanical work can be converted into heat and heat into work and the converted amounts of work and heat are equivalent to each other.

The Second Main Theorem of Thermodynamics (entropy theorem) specifies the direction of thermodynamic changes of state. According to this law, in nature only those processes (irreversible or natural) can run by themselves, in which entropy is exchanged by the system with the environment or produced in the system. This principle determines the direction of the process flow, and the entropy increase is a measure of the non-reversibility of a process.

The Third Main Theorem of Thermodynamics is also called the most fundamental statement that two systems which are in thermal equilibrium with a third system are also in thermal equilibrium with each other. From this follows the existence of temperature as a new, intensive state variable in addition to the mechanical quantities (pressure, volume), which is the same everywhere in equilibrium systems in Chap. 3.

When Georgescu-Roegen postulates a Fourth Law of Thermodynamics, this can only be met with the greatest scepticism. However, we must qualify that his interpretation of the law of energy is also highly controversial in the social sciences, "which is why many authors settle for an intuitive notion of entropy, which can result in more or less ambiguity" (Brunner 1997).

Summary

Security research should focus its work on transferring the entropy concept to the social contexts of decisions. This could make it possible to show that human misconduct does not endanger the natural basis of human beings.

Through the existing strict separation of natural science and social science, science loses its orienting power for a prosperous shaping of the social interrelationships of nature and of society itself.

References

Benett MR, Hacker P-MS (2006) Philosophie und Neurowissenschaft. In: Struma D (Hrsg) Philosophie und Neurowissenschaften. Suhrkamp, Frankfurt

Bräuer K (2005) Gewahrsein, Bewusstsein und Physik: Eine populärwissenschaftliche Darstellung fachübergreifender Zusammenhänge. Logos, Berlin

Bräuer K (2009) Das Weltbild der modernen Physik. www.kbraeuer.de. Accessed: 15. Dez. 2018

Brunner K-W (1997) Gesellschaftliche Integration der Umweltthematik: Zur Neustrukturierung einer Differenz. In: Rehberg K-S (Hrsg) Differenz und Integration. Westdeutscher Verlag, Opladen

Clausius R (1870) Ueber die Zurückführung des 2. Hauptsatzes der mechanischen Wärmetheorie auf allgemeine mechanische Prinzipien, zitiert aus Serge' E (1984) von fallenden Körpern zu den elektomagnetischen Wellen, Piper, München

Dahms K (1963) Über die Führung. Ernst Reinhardt, München

Falkenburg B (2012) Mythos Determinismus Wieviel erklärt uns die Hirnforschung? Springer, Heidelberg

Franken S (2007) Verhaltensorientierte Führung. Gabler, Wiesbaden

Goff P (2017) Consciousness and fundamental reality. Oxford University Press, Oxford

Hauptmanns U, Herttrich M, Werner W (1987) Technische Risiken. Springer, Berlin (Geleitwort von Prof. Dr. Klaus Töpfer)

Helfrich H (1996) Menschliche Zuverlässigkeit aus sozialpsychologischer Sicht. Z Psychol 204:75–96

Hochgerner J (2012) Publiziert in Thomas J und Sietz M. Nachhaltigkeit fassbar machen. Entropiezunahme als Maß für Nachhaltigkeit. Favorita Papers, Diplomatische Akademie Wien

Hoensch V (2006) Sicherheitsgerichtetes Leistungsverhalten in Kernkraftwerken. Utz, München

Janich P (2006) Der Streit der Welt- und Menschenbilder in der Hirnforschung. In: Struma D (Hrsg) Philosophie und Neurowissenschaften. Suhrkamp, Frankfurt

Lesch H (2016) Die Elemente, Naturphilosophie Relativitätstheorie & Quantenmechanik. uni auditorium. Komplett-Media, Grünwald

Luhmann N (2006) Organisation und Entscheidung. VS Verlag, Wiesbaden

Magic of Mountains. https://laemmchen.blog/2016/01/17/die-zeit-auf-dem-zauberberg/. Accessed: 6. März 2019

Mohrbach L (2012) Seebeben und Tsunami in Japan am 11. März 2011. VGB PowerTech, Essen

Müller K (1996) Allgemeine Systemtheorie. Springer, Wiesbaden

Nida-Rümelin J, Rath B, Schulenburg J (2012) Risikoethik. De Gruyter, Berlin

Ostwald W (1909) Energetische Grundlagen der Kulturwissenschaft. Philosophisch-soziologisce Bücherei, Verlag von Dr, Werner Klinkhardt, Leipzig

Penzlin H (2014) Das Phänomen Leben Grundfragen der Theoretischen Biologie. Springer, Heidelberg

Perrow C (1987) Normale Katastrophen: Die unvermeidlichen Risiken der Großtechnik. Campus, Frankfurt

Prigogine I, Stengers I (1993) Das Paradoxon der Zeit, Zeit. Chaos und Quanten, Piper, München

Rasmussen J (1986) Information processing and human machine interaction. North Holland, New York

Schuler H (1998) Organisationspsychologie. Huber, Bern

Schaub H, Störungen (2018) Fehler beim Denken und Problemlösen. https://www.psychologie. uni-heidelberg.de/ae/allg/enzykl_denken/Enz_09_Schaub.pdf. Accessed: 14. Dez. 2018

Searle JR (2006) Geist. Suhrkamp, Frankfurt

Sheldracke R (2015) Der Wissenschaftswahn Warum der Materialismus ausgedient hat. Droemer, München

Spanner G (2016) Das Geheimnis der Gravitationswellen. Kosmos, Stuttgart

Struma D (2006) Ausdruck von Freiheit. Über Neurowissenschaften und die menschliche Lebensform. In Struma D (Hrsg) Philosophie und Neurowissenschaften. Suhrkamp, Frankfurt, S. 187–214

Suppes P (1984) Probabilistic metaphysics. Blackwell, Oxford

The National Diet of Japan (2012) The official report of the Fukushima Nuclear Accident Independent Investigation Commission. Executive summary. https://en.wikipedia.org/ wiki/National_Diet_of_Japan_Fukushima_Nuclear_Accident_Independent_Investigation_ Commission. Accessed: 5. Dez. 2018

Tuss S, Bern AL (2013) Observer – induced quantum mechanical state collapse in the Libet experiment. J Exact Results Philos 1:1

Tsunami. https://de.wikipedia.org/wiki/Tsunami. Accessed: 14. Dez. 2018

Zitate von Albert Einstein: gutezitate.com/zitat/102158. Accessed: 14. Dez. 2018

Index

Printed in the United States
by Baker & Taylor Publisher Services